THE COMPLETE BOOK ON SPEED ENFORCEMENT

THE COMPLETE BOOK ON SPEED ENFORCEMENT

A Practical Guide to Understanding Speed Enforcement Concepts and Devices

By

KEVIN M. MORRISON

CHARLES C THOMAS • PUBLISHER, LTD.
Springfield • Illinois • U.S.A.

Published and Distributed Throughout the World by

CHARLES C THOMAS • PUBLISHER, LTD.
2600 South First Street
Springfield, Illinois 62704

ISBN 978-0-398-08733-3 (paper)
ISBN 978-0-398-08734-0 (ebook)

Library of Congress Catalog Card Number: 2011043654

With THOMAS BOOKS *careful attention is given to all details of manufacturing and design. It is the Publisher's desire to present books that are satisfactory as to their physical qualities and artistic possibilities and appropriate for their particular use.* THOMAS BOOKS *will be true to those laws of quality that assure a good name and good will.*

Printed in the United States of America
MM-R-3

Library of Congress Cataloging-in-Publication Data

Morrison, Kevin M., 1954–
The complete book on speed enforcement : a practical guide to understanding speed enforcement concepts and devices / by Kevin M. Morrison.
p. cm.
Includes bibliographical references and index.
ISBN 978-0-398-08733-3 (paper) -- ISBN 978-0-398-08734-0 (ebook)
1. Radar in speed limit enforcement--United States. 2. Radar in speed limit enforcement--Detection--United States. 3. Traffic regulations--United States. 4. Automobiles--Speed--Measurement. 5. Optical radar. 6. Speed limits. 7. Traffic police--Equipment and supplies. I. Title.

HV8079.5.M67 2012
363.25′947--dc23

2011043654

In Memory of Sergeant Robert McLane Hall
(1949–2011)

PREFACE

Two intentions motivated the writing of this speed enforcement book: to give the reader a thorough understanding of speed enforcement concepts along with the devices that measure speed and, by means of many interesting applications, to prepare them to use these concepts and devices in their own professional careers.

I have heard many concepts about speed enforcement over the years and ultimately ended up becoming a speed enforcement instructor. During the time as an instructor, I kept on learning about both the methodology and the technology. I discovered that there was always one more thing my students wanted to know. I also discovered that in each class I taught, every student gained from a different part of the material. This book was written to give complete knowledge to the reader so that they can take from it something new each time they read it.

The book starts with a basis for the concept of speed in scientific terms. Speed is often the word associated with going "too fast." In reality, speed is simply a measure of motion. The book will explore some concepts of speed as they relate to energy and work. Understanding the physics of motion will help the reader to better grasp the realities of speed as it relates to its effect on safety. Also, both the positive and negative effects that various speeds have on the outcome of travel will be explored. Speed is useful to determining time expended and fuel usage and each of these have a correlated cost consideration. These concepts set a positive outcome at certain speeds and conversely they have a negative outcome when improper speeds are applied. Because of these negative effects, limits are established by communities and government. This book also addresses the methodology to determining a speed limit. The scientific process of establishing speed limits along with the emotional aspect of limiting speed is covered in terms the reader can relate to.

Fundamental to this book are the devices used to measure the amount of speed. From the basic concepts of time distance determination through instant speed detection devices such as speed radars and speed lasers, the reader will explore the historical developments, scientific principles, and operational considerations of these devices. These sections are presented with in-depth information using charts and illustrations to assist in a complete understanding. The reader will be given this information in easy to understand language in order to have a full grasp of both the scientific and operational differences of each technology along with the operative limitations associated with these technologies.

If you are using this book as a text for speed enforcement classes, you will find it to be comprehensive concerning all methods used to monitor and enforce speed laws. If your scope for the training you plan to provide is less than the subjects in this book, you may apply it in four tracks that are most popular:

- For Time Distance Applications, it is recommended that you use Chapters 1, 2, 3, 4, and 10.
- For Radar Applications, add Chapters 5 and 6 to the Time Distance concentration.
- For Laser Applications, add Chapters 7 and 8 to the Time Distance concentration.
- For Radar and Laser combined, the entire book is recommended.

The overall goal of an effective speed enforcement program is to instill a desire to obey the speed laws, therefore resulting in fewer citations issued as fewer violations occur. The desired outcome is to get the public to obey the speed laws because they believe in them. This book explores the concept of voluntary compliance and the associated methods of achieving such compliance.

The fact that many people do not immediately subscribe to speed laws as being needed, directs that those citations which are written must withstand court scrutiny. As such, the book will also review the elements needed for court acceptance of both speed law violations and the instruments used to measure those speeds. Actual case examples are used to emphasize the major legal points.

Speed enforcement has been a public concern for over 100 years. Devices to determine speed thereby assisting law enforcement in the detection of those exceeding speed limits, are constantly transitioning to meet current needs. Therefore this book ends with an examination of some devices on the horizon.

An important note for the reader: This book is written with quantities and units in both U.S. and metric measure so that the text can be read in either system independently of the other.

K.M.M.

CONTENTS

THE COMPLETE BOOK ON SPEED ENFORCEMENT

Chapter 1

PHYSICS OF MOTION

People have a built-in sense concerning motion. The body detects motion and the mind interprets it. When riding a roller coaster, people sense the movements of the car they are riding in, but they relate those movements to past experiences. As the roller coaster car climbs the initial length of track moving towards the main drop, there is anticipation. Then as the car drifts across the top, descends, and accelerates, the body feels the fall and the mind identifies the fact that in this situation you will need to stop at the end; preferably a controlled stop!

If we were to freeze time halfway down the fall, and then ask the riders what they think is about to happen, most will tell you that the car they are riding in will arc out of the fall and continue on to the next stretch of track. Some may be concerned that the speed they are falling and the weight in the car will cause them to derail. With a very rare exception, none of them could predict the potential danger they may face. They have the reassurance of past rides or having watched other roller coasters and observed that they ended without harm.

Maybe if I explained the physics behind the movement of the roller coaster, I could assure them and instill confidence. If they understood Newton's laws of motion, and were given the formulas to compute how the car will perform in their given circumstance, they would rest easy on the ride. Although this would be somewhat helpful, it would be impractical. For one thing, they don't want to be reassured. They took this ride for the thrill and they certainly don't want this ride to be mundane. But, it's important to know that those individuals who designed this ride have a working knowledge of the physics of motion.

In this chapter, you will be introduced to some concepts in physics as they relate to motion. They are presented here so that you can have reassurance and confidence that the facts in this book have a basis in science and math. Join this with the concepts of how things work and top it off with some experience to test the theories and concepts expressed here and you will end up with a high level of knowledge in speed enforcement.

TIME AND DISTANCE IS SPEED

Almost everything in the world moves or has the potential to move. Probably the first studies in physics dealt with movement. This statement can be made since there is an extensive amount of study into this subject. Of the two individuals who stand out above others, Galileo Galilei (1564–1642) and Isaac Newton (1642–1727), it was the theories of Isaac Newton that resulted in a science discipline of its own, Newtonian Physics.

Speed is a measure of motion. Speed refers to how far an object moves in a given amount of time. How precise the measurement of speed is dependent on the amount of time and distance used for the calculation. Therefore, the general speed equation may be expressed as:

$$\textit{Average Speed} = \frac{\textit{distance traveled}}{\textit{time elapsed}}$$

When this formula is used, the units of distance and time should be expressed in a conventional way. We usually speak of things as traveling a certain number of *miles per hour* or *mph* (*kilomters per hour* or *km/h*). The word *per* is represented by the divisor in the general speed equation and means in relation to some specific unit of measurement, in this case, a unit of time. For vehicle speeds, the measurement is often stated in units of one hour (i.e., miles per hour or kilometers per hour). If we are traveling on a long road trip and we started counting the distance markers on the side of the highway, the total markers we passed in one hour would give us a measure of our speed based on the standard distance the markers represented.

As you read this book, you may have noticed I used the term speed in relation to feet per second (meters per second). Those readers who may reconstruct traffic crashes learned that the term feet per second (meters per second) is a determination of velocity. The fact is that in everyday life, we use these terms interchangeably. In physics, the term velocity is used to signify magnitude of how fast an object is moving *and* its direction. For example, direction can mean acceleration (a positive direction) or deceleration (a negative direction). This would be common in crash reconstruction. Speed, on the other hand, is only magnitude.

In this book, both miles per hour (kilometers per hour) and feet per second (meters per second) are used to solve for speed, since we are only addressing uniform motion or quantifying motion. Velocity is used where laws of physics were discussed since they were originally stated that way.

Since we are going to express speed as miles per hour or kilometers per hour, it is important to understand that this time and distance formula is foundational to all speed calculation regardless of the methods used to determine the speed. In fact, each measurement taken to determine speed can be derived from the general speed equation. Let's look at a GPS unit that gives your speed as you are traveling. In one instant, the GPS has located you at a point on the earth. In the next instant, you are at a new point. Since latitutde and longitude positions are actual fixed locations, a change from one to another represents motion, or distance traveled. And with knowledge of the amount of time elapsed, this motion can be expressed as speed.

There is an inverse rule to the general speed formula. Since speed is a result of the relationship between the distance traveled and the time elapsed, then the following must be true:

1. The greater the distance and the shorter the time, then the higher the speed.
2. The greater the time and the shorter the distance, then the lower the speed.

Table 1.
TIME AND DISTANCE RELATIONSHIP TO
SPEED COMPARISON (ENGLISH AND METRIC)

SPEED (KM/H)	*DISTANCE*	*TIME*
50.5	91.4 meters	6.0 seconds
58.4	97.5 meters	6.0 seconds
62.1	103.6 meters	6.0 seconds
65.8	109.7 meters	6.0 seconds
69.5	115.8 meters	6.0 seconds
73.0	121.9 meters	6.0 seconds

SPEED (KM/H)	*DISTANCE*	*TIME*
73.0	121.9 meters	6.0 seconds
67.4	121.9 meters	6.5 seconds
62.6	121.9 meters	7.0 seconds
58.4	121.9 meters	7.5 seconds
50.5	121.9 meters	8.0 seconds
51.6	121.9 meters	8.5 seconds

By now we can see that the way we calculate speed is dependent on the distance and time units we use. Units of measure for distances are typically expressed as miles (kilometers) or feet (meters). Time can also be expressed as hours, minutes, or seconds. For example, a distance of a quarter of a mile may be expressed as .25 miles or 1320 feet (a quarter of a kilometer would be .25 kilometers or 250 meters). In speed enforcement, we express a speed measurement in miles per hour (kilometers per hour) as the legal system defines tolerances in this manner. However, when dealing with small distances, such as 300 feet (91.44 meters), it is somewhat difficult to express. A vehicle that travels a distance of 300 feet (91.44 meters) and does so in 3.6 seconds can be expressed much easier as 83.3 *feet per second* (*25.4 m/sec*). This is convenient since our description of the distance is already in feet (meters) and our description of time is already expressed in seconds.

Therefore, when working with distances considerably less than a mile and time that is considerably less than an hour (or even a minute), we would use the convention of feet per second or fps (meters per second or m/sec). To change to an expression of feet per second from a speed of miles per hour or mph (kilometers per hour or km/h), we can use the general speed equation and plug in the known data.

Average Speed = distance traveled / elapsed Time[1]
U.S.
Speed = distance in feet / time in seconds
Speed = 300 feet / 3.6 seconds
Speed = 83.3 fps
Metric
Speed = distance in meters / time in seconds
Speed = 91.4 meters / 3.6 seconds
Speed = 25.38 m/sec

Now it is necessary to convert the expression from feet (meters) to miles (kilometers). The correct process is to take the answer from the equation we just solved and insert it into the speed conversion equation below:

Speed in mph = fps / (feet in a mile / seconds in an hour)
(Speed in km/h = m/sec / (meters in a kilometer / seconds in an hour))

As you can see we are converting from feet to miles (meters to kilometers) and seconds to hours in this equation. Let's now solve this equation:

U.S.
Speed in mph = fps / (feet in a mile / seconds in an hour)
Speed in mph = 83.3 fps / (5280 feet / 3600 seconds)
Speed in mph = 83.3 / 1.4666
Speed in mph = 56.8 mph
Metric
Speed in km/h = m/sec / (meters in a kilometer / seconds in an hour)
Speed in km/h = 25.38 m/sec / (1000 meters / 3600 seconds)
Speed in km/h = 25.38 / .2777
Speed in km/h = 91.4 km/h

1. Note: For consistency all formulas and calculations in this book, from this point forward will be displayed as Algebraic Expressions.

You should see that the conversion factor for changing a speed in fps to mph is 1.4666 and the conversion factor for changing m/sec to km/h is .2777 in the above equations. Both of these numbers are recurring decimals (decimals that never end). In these math problems the calculations were performed with some truncation (calculating to a certain number of decimal places and then ignoring all further recurring decimals). Even by leaving the numbers in raw form (i.e., 5280/3600) in the calculator, you will still have differences depending on the precision of the calculator. These differences will be minor, but may be noticeable. Of course, very few speed measurement devices show any decimal at all, and those that do will generally show only the first decimal place.

ENERGY AND SPEED

We can consider that it takes effort to move. On a large scale, a locomotive demonstrates this work theory very nicely. A locomotive at rest can weight over 400,000 pounds. Without going into the scientific reasons, most anyone can recognize the amount of effort it takes to move such a massive vehicle. This statement can also be expressed as work. The diesel engine must be capable of producing sufficient work to move the locomotive. When the locomotive is at rest, it is said to have static energy. But when the locomotive is in motion, it is said to have kinetic energy (kinetic meaning motion). Kinetic energy can help define the amount of work produced by a moving object. Consider the mathematical definition of kinetic energy as:

Kinetic Energy = 1/2 times mass times velocity squared ($KE = \frac{1}{2} mv^2$)

In this equation, *KE* represents Kinetic Energy, m represents the object's mass, and *v* represents the object's velocity.

The term mass and weight are often thought of as the same, but it is important to understand that mass is a property of the object while weight is a force, the force of gravity acting on the object. For now, understand that mass is a constant for the object; it does not change. Velocity which represents its motion does change. Let's look at an example. Take two objects that are the same mass, one of those objects moving at twice the velocity as the other. The faster moving object will have greater kinetic energy than the slower one. In other words, there is more work energy in the faster object than in the slower one. A hammer has more mass than a nail. When the hammer collides with the nail, a force is exerted on the nail and the nail moves a distance. So the amount of force (work) in the collision is dependent on the amount of kinetic energy.

We have stated that the mass is constant and therefore the work, or force, changes as the velocity (speed) of the object changes. Note that in the equation, velocity is squared. So if the mass is doubled, the kinetic energy is doubled; but if the velocity is doubled, the object has four times the kinetic energy and is therefore capable of doing four times more work.

Now let's apply this concept to speed enforcement. Law enforcement officers tend to set an arbitrary threshold to the speed at which they would stop a vehicle. Some officers may allow a vehicle a certain measure of speed over an established limit. There are many emotional reasons for having a threshold and the amount they allow. Some of these reasons are:

1. The average speedometers on cars are not very accurate.
2. The officer believes that he/she sometimes exceeds the speed limit and doesn't want to be a hypocrite.
3. The officer believes he/she can't stop them all, so they favor searches for the fastest vehicles.

Often the allowed speed given by most police officers is 12 mph (about 20 km/h). In a posted 30 mph zone (about 50 km/h), this threshold would double the kinetic energy of the vehicle. If we recall the example of the hammer, it can be readily seen that the damage potential of a motor vehicle increases geometrically with its speed. A motorcycle traveling at high speed can have the same energy, the same force, the same destructive nature as a car. Law Enforcement measures the severity of a traffic crash based on the destructive nature of that crash. In Figure 1, we can see the resulting damage from a high-speed impact. The vehicle pictured sustained heavy damage (and a fatality injury) from an impact estimated at 100 mph (161 km/h).

Figure 1. Damage resulting from a high-speed crash.

There are many factors that cause the resulting destruction, but based on what we see in this formula, if we increase the speed of a vehicle, we also increase the energy and therefore the destruction. Also, because we are dealing with the work the vehicle is doing, we must consider the effect that it has on the roadway it is traveling on. Isaac Newton's third law (put simply) indicates *that for every action there is an equal and opposite reaction.* If you push a door open, there will be resistance due to this law. If the door is locked, then the force applied to the door by pushing it is met with an equal and opposite force pushing back. As a vehicle travels on the roadway, the tires rotate and push against the roadway. The more speed applied, the more force exerted onto the roadway. This can be observed when driving on a muddy road. If a steady slow rate is applied, the tires create a slight rut in the road. But traveling faster will deepen the ruts. The same is true for a hard road.

The major problem with these concepts is that the driver of a speeding vehicle has a difficult time understanding or even believing these concepts. In Clearwater, Florida, engineers determined that the bridge which connected Clearwater Beach to Sand Key was in great need of repair. Although not yet unsafe to crossing vehicle traffic, the engineers calculated that by slowing the traffic to 15 mph (24 km/h), they could buy time until the repairs could be made. They restricted the bridge so that trucks were not allowed, and police officers from the Clearwater Police Department were detailed to run radar on the bridge enforcing the speed limit of 15 mph (24 km/h). The motoring public regarded the move as a way to pay for the bridge repairs. They did not believe that driving too fast over the bridge would have any detrimental effects. After all, rubber tires can't damage asphalt. And besides, isn't it better to minimize the time on the bridge. The bridge was finally repaired, but the officers who enforced the speed limit faced many irate citizens. Without the knowledge of the physics of motion, their job of attempting to convince the drivers of why they needed to drive slower would have been that much harder for them.

Finally, energy can be transferred. As a vehicle travels at a certain speed, so do the occupants. The effects of acceleration and deceleration on the vehicle are also experienced by those riding in the vehicle. As a vehicle travels at a steady speed, so does each part of the vehicle, as well as the driver's body and organs. Newton's first law of motion states that *a body continues in its state of rest or of uniform speed in a straight line unless it is compelled to change that state by forces acting on it.* This is referred to as inertia. When a vehicle strikes another object, it creates a changing force. If one vehicle is faster than another, the slower vehicle experiences the most change. The energy from the faster vehicle is transferred to the slower one. Like the hammer hitting the nail, the more speed in the swing, the more the nail moves.

But unlike the nail, a vehicle is made up of multiple parts held together by bolts and welding points. As a vehicle experiences change, the energy is transferred through the vehicle parts. The degree to which each part reacts to this change depends on the intensity, composition, and assembly. Steel parts may bend, while plastic parts may shatter. One of the parts that will ultimately experience this energy would be the occupants of the vehicle. Once the force of change reaches the occupants of the vehicle, injuries may occur. The speed of the vehicle and the degree of change are paramount to the forces that will affect the human body. When a vehicle moving at 60 mph (100 km/h) strikes an object and is suddenly decelerated, each corresponding part of the vehicle must also be decelerated from that speed. If the energy in the collision is not dissipated, the fragile parts of the human body will by subjected to this force, often to a level they cannot withstand. In the crash that took Dale Earnhardt's life at Daytona International Speedway in 2001, the car he was traveling in was moving at approximately 159 mph (255.8 km/h) when it encountered a concrete wall. The car changed speed due to the solid wall. The change in speed was estimated at 42 to 45 mph (67 to 72 km/h) which resulted in blunt trauma to the brain.[1] The race car lost this speed (42 to 45 mph) very quickly. Watching this impact on television, people did not regard it as extreme. How could someone die at such a low impact speed? But people can die from impacts at very low speeds. The human body is very fragile and even a head striking the pavement at 10 mph (16 km/h) can cause irreparable damage. What is extreme is the amount of speed change the body experiences in a very brief time.

In this chapter, we have seen that speed is a measure of movement and we have learned how to use that measure to determine speed and the effect on a moving object such as a vehicle. We have also learned that movement involves energy, which is called kinetic energy and that energy can be measured. More importantly, we have found that the greater the speed, the greater the potential for undesirable effects on vehicles, roadways, and human beings unless that speed is controlled.

---

1. Earnhardt Report as reported by Nascar.com, August 21, 2001.

Chapter 2

EFFECTS OF SPEED

In Chapter 1 we learned that speed is a measure of motion and that it is defined as the distance traveled divided by the time elapsed. Mathematically, speed is the solution of the two variables: distance and time. But, the average driver does not view speed in this way. The driver is concerned with getting there and so the driver sees the solution as being either distance or time. In the driver's eyes, speed becomes the variable. In a non-mathematical sense, many drivers see that "the faster I go, the sooner I will arrive." In fact, some drivers will rationalize, "I've got a lot of distance to cover so I'll have to get moving," as if the fact that if they go faster it will reduce the distance needed to travel. I realize that this makes little sense, but somehow it's the common justification to the speed they select to drive.

SPEED AND TRAVEL TIME

The amount of time it takes to go from point A to point B is dependent upon the speed traveled over a certain distance given the highway system used. Remember that in Chapter 1, we identified that speed had an inverse relationship to time and distance. Now we can see the same kind of relationship with time. Let us say you have a 10-mile (16 kilometers) trip, and the speed limit for the route you have chosen is 60 mph (100 km/h). This dictates that it will take you 10 minutes to drive that distance. But you need to get there sooner, so you decide to travel a little bit faster, and you step up your speed to 70 mph (112 km/h). By doing so, you should arrive at your destination 1 minute and 26 seconds sooner. It doesn't seem worth the effort does it? How about over a longer distance? Let us change this to a 60-mile (96 km) trip. Now increasing the speed by only 10 mph (16 km/h) more, you will arrive at your destination about eight minutes and 36 seconds sooner. What will you do with that extra eight minutes and 36 seconds? Maybe that will hold some value to you. But keep in mind we have examined the time it takes to travel in a sterile environment. There are a multitude of other factors affecting this travel time: traffic signals, other traffic, distractions, weather, and so on. Generally, the driver is only interested in one thing, which is the goal. If the driver left for work late and wants to make up the time so that his/her arrival is not late, then the justification is to speed up and try to "beat the clock." But the fact is that the distance to be traveled is a fixed number and the variables we have mentioned have the potential to rob the driver of the conceived advantage to speeding. Of course, there are other disadvantages to going too fast that we haven't discussed yet.

But what is too fast? Speeding is a relative term and depends on who is defining it and the environment it is established in. To a race car driver, the concept of going too fast means that the car exceeds the physical limitations of the track it is being driven on. To the homeowner standing in the front yard and watching the car traveling past his/her house, too fast isn't based on a number but more of a feeling. This person will call the law enforcement agency to let them know they have a "speeding problem." The officer responds and may or may not agree with the citizen that cars are traveling too fast in the neighborhood, only to be redirected by a supervisor who weighs the speeding problem against other priorities in the community. To that supervisor, the speed may not be too fast when compared to the burglary or theft problems of that patrol area. So one can see that the concept of too fast transcends the clear definition of simply distance traveled and elapsed time; emotional elements are also included.

DOES SPEED CAUSE TRAFFIC CRASHES?

We have set the stage so-to-speak for the potential that greater speed increases certain risks to people. Without directly stating it, you can see those potentials in the physics of speed. One thing we haven't looked at is the effect traveling faster has on the driver. So let us examine how speed may affect the driver.

Drivers come in a great variety. They may be young, middle age, or old, and they may have very little experience or they may have years of experience. One thing I have discovered is that regardless of the driver's experience levels, there are advantages and disadvantages to each. When I was learning to fly an airplane, I commented to my flight instructor that they must enjoy a student pilot who had military flight experience. My instructor told me that this was not always an advantage. Military pilots can handle extreme situations better, but they also have embedded bad habits. "Everyone performs based on their experiences, but the playing field is level when a pilot is experiencing something for the first time." This concept is true with drivers. A driver who is used to driving the autobahn in Germany may not do so well at high speeds on island roads in American Samoa. It becomes very dependent on experience in a particular environment. And likewise, familiarity with an environment causes some drivers to relax and become careless.

Certainty when traveling faster (without the qualification of too fast), the driver's abilities are taxed more than at slower speeds. When traveling at a speed of 35 mph (56 km/h), the driver will have a reaction time of (on average) 1.5 seconds. Driving an increased speed of 50 mph (80 km/h), this driver still has the same reaction time but will cover a greater distance during this time. If the driver fails to adjust for the greater distance that will be traveled at greater speeds, the potential for a traffic crash will increase. Let us break down the steps that occur when a problem like a child running into the road is faced by a driver.

> Step 1–Identifying. If the driver is paying attention, he/she will have to recognize that a problem exists. Of course, the faster you go, the less time you have to identify that the event you are approaching is in fact a problem.
>
> Step 2–Predicting. If the driver has the experience to conclude what will happen if action is not taken, this predicting can occur very quickly. If he/she lacks the experience, then this portion of the process will be longer.

Step 3–Deciding. Often there are several things that can be done to avoid the problem: steering away or braking, for example. Much of this step is dependent on the drivers conditioning.

Step 4–Executing. The driver then must put the decision into motion. Assuming that the driver does not "freeze up" and do nothing, the execution step time will depend on what was decided.

Steps one to three occur during what is called reaction time: the time from starting to identify the problem to the end of the time in deciding what to do. Engineers have placed a value on this reaction time through various testing methods. Initial testing sets this reaction time to three-quarters of a second (.75 seconds), but this time was disregarded later because the testing that was done was to an anticipated problem (in other words, the test subject knew that a problem was about to occur, which allowed the subject to be better prepared for the outcome). The final time settled on by most engineers is approximately one and one-half seconds (1.5 seconds). This time is affected by a variety of other factors such as sight distance, driver fatigue or impairment, and/or lighting. However, for this discussion, we can compare the different distances traveled at various speeds with a uniform reaction time. Also for this discussion, we will use a stopping distance based on a uniform frictional force for each speed. Stopping distances behave geometrically to the overall distance to stop than do reaction times.

To best understand these concepts, let us look at a case study. Throughout this study, you may refer to the reconstruction diagram shown in Figure 2.

The location is a straight two-lane undivided roadway. The roadway was made of asphalt and well maintained. The posted speed limit is 35 mph (55 km/h)[1] and traffic was very light with good visibility. Adjacent to this roadway is a field which is a common location for four-wheel recreation vehicles otherwise known as four-wheelers. Two young people were riding a four-wheeler in the field: a 13-year-old boy operating the four-wheeler, and a 12-year-old girl passenger. A Lincoln was traveling east on the road approaching the field. The Lincoln

1. At times the metric conversion will be adjusted slightly. In this case, the correct conversion is 54 km/h, but since we are speaking about a speed limit, it is far more likely that the speed would be posted at 55 km/h, not 54 km/h.

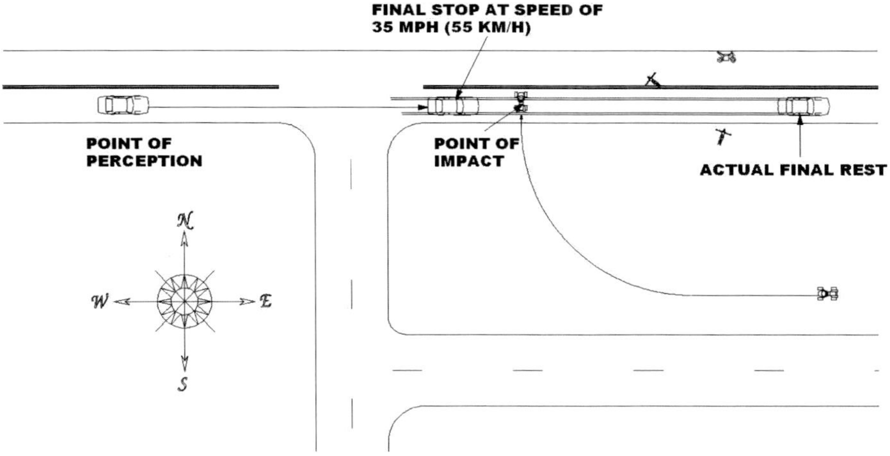

Figure 2. Reconstruction diagram of crash involving ATV and Lincoln.

was traveling 47 mph (75 km/h). As the Lincoln approached the field, the four-wheeler started to arc away from the roadway. The four-wheeler was traveling west in the field, when it made an abrupt turn towards the road. The driver of the Lincoln saw the turn and realized the path would take the four-wheeler onto the road and directly in front of the car. The driver reacted and applied full brakes. The Lincoln traveled a total of 73 feet (22.25 meters) in braking to a stop. The four-wheeler was struck 35 feet (10.6 meters) into the braking. The children were ejected and landed in the road ahead of the Lincoln's final rest position. Both children had severe injuries but did survive the crash.

Any investigator would identify that the four-wheeler cutting into the path of the car as being the cause of the crash. But what role did the speed of the Lincoln play in this event? One may believe that the speed intensified the severity of the crash, and they would be right. But can the speed of the Lincoln be considered a proximate cause[2] of the crash? Let us look at the reaction time based on the speed of the Lincoln.

With the Lincoln traveling at 47 mph (75 km/h) and with a reaction time of 1.5 seconds, the start of reaction time would be 103.4 feet

2. *Black's Law Dictionary* defines Proximate Cause as "That which, in a natural and continuous sequence, unbroken by any sufficient cause, produces the injury, and *without which the result would not have occurred.*"

(31 meters) from the point that braking began. That places the Lincoln 138.4 feet (41.5 meters) before the impact point. If the Lincoln was traveling the speed limit of 35 mph (55 km/h) and all other facts in this event, including the point of driver perception of the hazard, remains the same, the Lincoln would have traveled 76.9 feet (22.9 meters) or 26.5 feet (8 meters) less than the higher speed. Also the braking distance would have been 40.83 feet (12.44 meters) for a total distance traveled of 117.73 feet (35.34 meters) with reaction time and braking combined. Remember, the impact point was at 138.4 feet (41.5 meters) from the first point of perception. But if the Lincoln had been going 12 mph (20 km/h) slower (or at the speed limit), then the driver would have stopped the vehicle 20.6 feet (6.274 meters) *before* the impact point, and the crash would have never occurred.

One must take care when applying this logic to a highway event. Regard must be given to what people would do normally in a given situation. Law Enforcement is charged with examining the facts and determining whether or not a law was violated. Obviously, if the Lincoln had been going under the speed limit, then braking may not even have been involved. This starts to tread on a question that is more civil in nature. But then what could law enforcement do to prevent this crash from occurring? If patrolling the area, an officer observed the four-wheeler operating near the road, the officer could have stopped and warned the operator of the potential hazard. This is certainly in the scope of the officer's duties, and a good police officer probably would take such action. But, is there anything preventive that could be done concerning the car? The fact that the car was speeding (going too fast) according to the law *and* that speed was a factor in this crash shows that speed enforcement could also have been a measure of prevention. Independently, either action on the part of law enforcement would be good, but combined, there is a high probability that the crash would never occur.

SPEED AND FUEL CONSUMPTION

It is understood that the faster a vehicle is driven the more energy it requires. Therefore, more fuel will be used. At least that's something I realize when I do physical work around my ranch. This is also true for vehicles. If the engine is turning more RPM's (revolutions per minute), then it must be gobbling up the fuel to do so. When a car is purchased, the buyer generally examines the fuel economy decal to see what the "average" mile per gallon rating is, highway and city. What they are looking at is the average and therefore such a rating can be affected by other factors. In 2009, a study was done by Consumer Reports[3] in which a variety of vehicles were tested for their fuel economy at different speeds; it was concluded that most vehicles get their best highway fuel economy at 55 mph (88 km/h). In 2011, Consumer Reports republished this article online.

The most famous fuel economy effort with relationship to speed was the United States National Speed Law that set the national maximum speed limit to 55 mph (90 km/h) in 1974. This action was the result of the Emergency Highway Energy Conservation Act that prohibited speed limits higher than 55 mph (90 km/h). This action was politically motivated and was enacted to conserve fuel by having drivers operate their vehicles at optimum speed for fuel economy. It did require heavy enforcement efforts for all states as the motoring public did not agree with the limits[4] being imposed. This law was in effect for 14 years and in 1987, the U.S. federal government began to eliminate the restrictions. There was a perceived side benefit, however, in the saving of lives from this action. In the first year, the highway death toll dropped by about 4,000 lives over the previous year. Although the life-saving results were real, the factors by which those lives were saved are difficult to assign. On one hand, vehicles were being driven slower which statistically lowers fatality rates. On the other hand, heavy enforcement campaigns gave an ominous law enforcement presence on the highway. It is unclear whether either or both factors combined and influenced the lower death toll.

MANAGING SPEED

This section is directed to those who authorize speed enforcement: the managers in a law enforcement agency. Some in police management would express concern over officers who actively enforce speed

3. "Tested: Speed vs. fuel economy," Consumer Reports, September 2009.
4. See Chapter 3 on Speed Limits.

laws. They picture the officer hiding in the bushes or behind the billboard, wasting valuable time when they should be on patrol. I must agree that if not done properly, speed enforcement can lack efficient use of manpower. Most people do exceed the speed limit and as such, officers must apply some discretion to their enforcement practices or they would be stopping cars all the time. The common way to regulate who gets stopped and who gets a citation is to set a limit as to how many miles per hour (kilometers per hour) a vehicle must travel before a stop is made or a citation is issued. In the 13 years that I was a sergeant, I do not remember anyone commending an officer or our agency for sitting along the road with a radar unit for hours on end. Supervisors lose faith in an enforcement program when they see an officer in one location for hours and then find that the officer only turned in three citations.

The city I worked in had a major U.S. highway that traveled through it. This highway was the primary way for people to move north and south through the county and connect motorists to the largest cities in the county. The estimate was that 75,000 to 100,000 vehicles traveled this road daily. The chances of finding a speeding vehicle on this road were very good. But the number of crashes was low due to the roads limited access design. I personally issued several traffic citations for speed in excess of 100 mph (160 km/h). Officers would argue that there were very few crashes because of their enforcement efforts, but management would see it as a moot point. Before you start any speed enforcement campaign, there are some rules that must be followed in order to have success.

A speed enforcement effort cannot be the sole component of an effective traffic safety program. Traffic safety programs must be based on the three E's of traffic enforcement: Engineering, Education, and Enforcement. Take, for example, a road that seems well built and properly posted. If there are high incidents of traffic crashes, or a number of citizen complaints, possibly there is the need for a reassessment of the limits set for this road.[5] Or maybe a better educational campaign needs to be conducted to make the users of the road aware of the need for compliance. Finally, these considerations may be fine, and it may simply be that increased enforcement is needed. A program to push

5. See Chapter 3 for more information on setting speed limits.

compliance with speed limits will be ineffective without enforcement. Operating a speed enforcement program with this approach is imperative to public as well as agency acceptance.

In a traffic safety program the desired goal is to get voluntary compliance for the traffic laws. Probably, the reason why speeding laws are so controversial is because so many people disagree with them. The desire is to get the motorists to obey the speed laws because they know it's right. As a child, you were probably told not to do something that could hurt you. For one reason or another, you go ahead and do what you were told not to and you don't get hurt. Then you do it again. Each time you do it, you are being reinforced that what you were told was wrong, and your judgment is better than theirs. That is until you get hurt (of course, you don't let on that they were right until a few years have passed). So how do we get the public to obey the traffic laws and keep the supervisors and management from thinking we are wasting time? The best way is to follow some simple guidelines:

- Guideline 1–Be sure that your speed enforcement has a basis. You should be able to articulate that you were enforcing speed laws for one of three main reasons: due to an identified crash problem, in response to citizen's complaint(s), or a speeding violation that you personally observed and concluded that such enforcement was warranted. All of these reasons will involve your doing some background work. You may have worked a crash and think "we seem to have a lot of crashes here." Or a citizen tells you, "I can never get out onto that road because the cars are going too fast." These are excellent examples of a reason to conduct speed enforcement. Please note that speed enforcement should never be used to placate a supervisor or to meet some form of quota. When I was a patrol officer, I had another officer call me almost daily at 10:30 a.m. to meet him so that he could give me his handheld radar to use the rest of the day. I finally asked him why he did this and he told me, "the sergeant needs two tickets from us a day, and I've got my two."
- Guideline 2–Don't pick a spot and sit for hours on end. This is like wanting to catch that elusive bass in the lake. You will wait hours to catch him. Instead, use a hit and run approach. If you can use a speed measuring device that allows you to continue to

patrol, that would be best (there are several identified in this book). If you have to sit, sit for a period of time and then patrol some and come back. There are methods to determine the best times to be at certain locations. We will explore those methods in the next chapter. Of course, there are some priority issues with this guideline. Often the best times are also the high activity times which will take priority. An effort to catch a speeder is more appropriately expressed as an effort to prevent loss of property or life. But the act to cause this loss has not happened and may not happen and therefore calls for service have priority over speed enforcement efforts. If you were patrolling the alley behind some businesses on midnight shift, and a call of a fight in progress came in, you would not continue to patrol the rest of the businesses before you took the call.

- Guideline 3–There is a lot of debate over covert enforcement, but it has been long established that visibility does a great deal of good in impacting the most people. In 1972, an experimental study was conducted in Kansas City, Kansas, when marked and unmarked police cars where used to determine if there was a reduction in crime based on visible patrolling. The outcome of the study determined that there was no appreciable reduction in crime. What the study failed to consider was that in order to respond to all calls for service, marked police cars had to drive through the zone that were assigned to only have unmarked cars. In essence, the study was inconclusive. Despite this study's failure, law enforcement everywhere is aware of the unscientific fact that motorists tend to slow down upon seeing flashing lights on a police car; something commonly referred to as "blue light fever." A quick look at Figure 3 and your first thought is a traffic violation. A person driving by an officer who is positioned with the emergency lights activated while behind a car will slow down with no real knowledge of what is going on. It could be something as simple as a disabled vehicle, but the motorist immediately assumes that the driver is getting a citation. There is nothing wrong with covert positioning of the patrol car or driving a patrol car that is not marked, but all stops should be as visible as safely possible.

Figure 3. An officer on the side of the road conducting a traffic stop.

- Guideline 4–Finally, have a method to measure success in your program and report these successes. This is not in the number of citations issued, although it will become a part of the report. High numbers of citations do not necessarily curb crashes. According to the guidelines mentioned above, there needs to be a purpose and therefore there needs to be a solution. There is more victory in goodwill between law enforcement and the public. Compliance is desired, and when it is achieved, report such a success to the agency and the public. And in speed campaigns, it is important to measure success in injury and death rates. A 10 percent drop in all crashes can be attributed to many things: changes in economy, public events, and demographics, for example. But a 10 percent drop in crash injuries does lend itself to effective speed enforcement, and this should be announced to the public. When they see the success, they too will buy in. And don't expect that the success will last. Additional enforcement will be needed from time to time. If conducted properly, these additional efforts will be minimal when compared to the program that was run to achieve compliance.

Postscript: To all enforcement, but especially speed enforcement, do not allow your officers to become too focused. Keep their eyes open to the other things that may be going on. Our senses often are dulled by the repetitive confrontation of dealing with so many speeders. Keep watch for other crimes that are detected through good speed enforcement. You are pulling over people who may have just been involved in a crime, or possibly are committing or about to commit a crime. Your stop of a speeder can be the doorway into a larger case than just a speeding citation.

Chapter 3

SETTING THE LIMITS

There is always someone who thinks "that's just too fast" for almost anything that moves. In 1830, a man from New York invented the steam locomotive named the "Tom Thumb" which could travel 18 mph (29 km/h) in an effort to go faster than the barges that traveled the Erie Canal. By the end of the 1800s some steam locomotives could travel an average of 50 mph (80 km/h) and some reached speeds of 70 mph (120 km/h). Yet people of that time felt this was too fast. Finally, someone would say, "There should be a law against that" and, of course, there soon was one. But when was the first automobile speed limit and how are they set anyway? In the United States, the state of Connecticut lays claim to the first automobile law which happened to be a speed limit law, established in 1903. It was a violation of this law for any vehicle to travel in excess of 12 mph (19 km/h). What method was used to determine this particular speed is unclear. As we passed from the Model T to today, legislatures have used several methods to determine speed limits and the resulting laws can be assigned to three different kinds of speed laws.

SPEED LAWS: THEN AND NOW

The first type of speed law is called the **Basic Speed Law**. This law is fundamental to the establishment of all speed laws, but it can be considered subjective because there are few absolutes. Many locations have a different variation of this law, but, in general, it states that drivers may not operate vehicles faster than is considered *reasonable and prudent* for the conditions. But what is reasonable? What is considered prudent? That is determined by others, possibly those in legal authority to make the law. But every person may interpret the speed traveled differently. For example, a motorist driving in an unpopulated area, on a straight road with no intersections and fences on both sides of the road may feel that a speed of 120 mph (193 km/h) would be reasonable. But the officer who stops that vehicle may feel it is neither reasonable, nor prudent to travel the speed he observed and therefore a citation is issued. Then in court, the judge hearing the case may agree with the motorist and find a verdict of not guilty. Conversely, a motorist driving the autobahn in Europe may feel that the speeds are unreasonable because they appear faster than comfortable.

The next speed limit to emerge was the **Prima Facie Speed Law**. These laws were enacted to give guidance to basic speed laws. In many instances, it established conditions by which the motorist must exercise caution. Such laws are still in effect today in states such as California and Florida. In some jurisdictions the wording of the law may even give the motorist reason to defend their speed. For example, a motorist driving a road may feel it is safe to increase his/her speed. The driver can reason that the highway conditions changed and therefore the vehicle may be driven at a higher speed. In understanding this type of speed law, keep in mind that Prima Facie means "at first look." It is designed to allow for flexibility, but like the Basic Speed Law, is open to individual interpretation.

Today most speed limits are set as **Absolute Speed Laws**. These laws are far more objective. The limit is established by statute and identified by signage informing the motorist of the limit. It is absolute and there is no room for interpretation (although there is still some interpretation in the application and adjudication of a speeding charge). With absolute speeds, a scientific process is used to determine the limit for a particular roadway, and the motorists are expected to comply. Speed limits that are set by this scientific method generalize the potentials for a traffic crash without regard for the human factor involved. The limit encompasses every level of driving experience from the race car driver to the young person driving alone for the first time.

THE SPEED STUDY

As I indicated, today's speed limits have their foundations in the Basic Speed Law. Consider this excerpt from the Michigan Motor Vehicle Code:

> The basis for all speed controls is predicated upon the nationally accepted principle that the majority of drivers are cautious, prudent, and drive at speeds that are reasonable and proper regardless of the posted speed limit.[1]

As you can see, there is an assumption that motorist will voluntarily drive at a reasonable and prudent speed. The speed that people drive all over the world is based in their knowledge of the highway they are operating on, and the familiarity with the vehicle they are operating. A driver who is traveling a mountain road in a car they have owned for some time will travel this road faster than a driver who is on the same mountain road but does not live in the mountains and is driving a rental car. In the vast majority of cases, there are more people driving this road who have experience driving it than the number of those who are just visitors.

The 85th Percentile Speed

This is the premise for the speed study. The engineer conducting the study will record as many speeds as possible to establish a statistical curve. If the road was designed correctly, and the study is conducted correctly, the outcome should be a normal bell curve. The desire is to determine the **85th percentile** for the road. The engineer will conduct this study using a method by which each vehicle traveling the road is counted and their speed is determined and recorded. This may be done by placing sensors in the road to record this information.

It is sometimes done with radar or laser speed studies; however, this method is labor intensive. The engineer needs at least a full week or more of data to get a sufficient population for the study. Manual methods are more of a sampling and can have inconsistencies. At

1. Michigan State Law Section 257.627.

Figure 4: Speed study by two college students.

times, the engineer actually collects the data, and at other times, it may be left to others such as college students or even law enforcement officers as shown in Figure 4. Once the data is collected, the engineer will determine the 85th percentile.

The best speed determination is completed using the data collected and calculating the speed at which 85 percent of the vehicles are traveling. Entering the data into a table, we can view how to arrive at this number. Observe that in Table 2, the first column is the speed at which the cars were traveling, while the second column shows how many cars were traveling at that speed. In this example, a total of 100 vehicles were recorded. In simple terms, the total number of vehicles multiplied by .85 will give the vehicle at the 85th percentile. In this case, that vehicle is the eighty-fifth from the slowest vehicle to the fastest (see the cumulative frequency column in Table 2). The 85th percentile vehicle was traveling at 36 mph (58 km/h). Since it is unnatural to set a speed limit at that speed, a range is used to establish the 85th

Table 2.
AN ACTUAL SPEED COUNT AND 85th PERCENTILE DETERMINATION

Speed (mph)	Frequency of Vehicles	Cumulative Frequency	Cumulative Percent	Speed Percentile
15	1	1	1%	
18	2	3	3%	
21	6	9	9%	
24	12	21	21%	
27	13	34	34%	50%
30	20	54	54%	
33	18	72	72%	85th
36	14	86	86%	
39	6	92	92%	
42	6	98	98%	
45	1	99	99%	
48	1	100	100%	

percentile.[2] As you can see in Table 2, that speed is between 33 and 36 mph (53 to 58 km/h). The engineer would then recommend a speed that is common for the jurisdiction where the limit is to be set. In this case, the recommended speed would be 35 mph (in metric the speed limit would be set at 55 km/h even though the conversion of 35 mph is 56 km/h).

Keep in mind that this was only 100 vehicles and would not be considered sufficient for the establishment of a speed limit. The larger the population, the more exact the 85th percentile can be determined, and additional formulas are used to verify the finding (i.e., standard deviation, etc.). The final speed limit may be based on this process but there will be adjustments for other considerations such as potential hazards in the area. Also keep in mind that a road design may support

2. Not all engineers will use a range such as this and there is a bit more complexity to this process. This example is provided for understanding.

a higher speed, but design will only affect the upper speed allowed, and not the actual speed limit the road is set to, which is often lower than the road-designed speed.

Emotional Speed Limits

The engineer will submit the recommendation, and on adoption, signs will be placed. But this adoption has other factors that I like to refer to as emotional considerations. The engineer is guided by something called warrants. This is not the kind of warrants that law enforcement is used to. These warrants are a determination of whether or not certain action is warranted. In the speed study example we just discussed, the location where the limit is to be placed may have a school or park adjacent to it. The engineers must determine if the limit should be lowered due to these factors. Often these considerations are addressed in a manual or guide book of warrants. In the United States the book most commonly used is the *Manual on Uniform Traffic Control Devices* (MUTCD), which is published by the U.S. Federal Highway Administration. In Europe and Asia, there are similar manuals and hundreds of studies which aid their engineers as well.

I have referred to this as the Emotional Speed Limits because they are not based in science, but are moreover based on emotion. An example of this may be found in the setting of school zone speeds. The school zone speed is contrary in nature to the driving habits of motorists. We cannot use the 85th percentile to establish a school zone. Without enforcement many motorists would simply ignore the speed limit and drive as usual. Motorists must be forced to reduce their speed for the safety of the children. If child safety was the concern of every driver, then any time a child is observed near a road, motorists would alter their speed. It is not that some don't do this, but it is against the nature of drivers because traveling slower makes them feel uncomfortable. Any time the driver is forced to do something he/she doesn't really want to do, it becomes an emotional issue. Construction or work zones are another emotional speed limit. It is not that the basis for the reduction in speed fails to have support in statistics, but rather it is the general way that it may be applied. Included in the emotional speed limits are those locations that did not do a study at all.

A neighborhood may have a series of local streets and the jurisdiction that is responsible for this community may not have a traffic engineer. So the limits are placed based on what is common for such communities without a study to back it up. Is this wrong? Hardly, since in the absence of an absolute speed law, the prima facie or basic speed laws may prevail. Also, generally, there is an absolute speed law that the community fathers adopted to control those locations where limits are needed but signage is not in place. The law may read as follows: "when no other speed limit is established, vehicles traveling within the limits of (jurisdiction name) will not exceed (some determined speed)."

Finally, it needs to be understood that roads that have established speed limits are subject to change. A speed study does not last forever. Changes in road designs and changes in traffic patterns do occur regularly. Allow me to give you an example.

A speed study had been done for a road that was a two-lane blacktop, with a single curve at one end. The posted limit was 30 mph (50 km/h) and the area was predominately residential with an elementary school next to it. The road had become a major artery bypass (this road paralleled another major artery road) and traffic volume had increased so the road was improved. Among the improvements was road widening to six lanes divided by a 20-foot grass median. The school was fenced off from the road and a flashing crosswalk signal was placed in front of the school. In addition, the curve was banked for speeds up to 65 mph (104 km/h). Once completed, the construction company reinstalled the 30 mph (50 km/h) speed limit signs. The reason for the original speed was because they did not have any additional speed study data with which to justify a change in the speeds.

This resulted in a natural tendency for the motorists using this road to drive at a more comfortable (higher) speed. The information originally gathered was no longer valid. Ultimately, a study was done and the limit was raised to 40 mph (65 km/h).

SPEED TRAPS

I always cringed whenever someone referred to speed enforcement as a speed trap. Generally, this term was used by the media and others who did not understand the significance of effective speed enforcement. A speed trap is thought of by some supervisors as a "fish-ing hole." This is a place officers can go to get citations when the agency tries to measure performance by the number of citations issued. There are three main causes for these locations where speeding is commonplace:

1. Speed limits that are set lower for legitimate reasons such as road construction.
2. Speed limits which are arbitrarily too low (as in the above example of the needed speed study).
3. Roads containing traffic volume that is significantly higher than most.

The first cause demands a concerted effort on the part of law enforcement to promote the need for motorists to obey the speed laws, as we discussed in Chapter 2. The second cause requires that law enforcement adjust their activity. Since officers are not engineers and they do not have speed study data to justify their enforcement, they must rely on those engineers to assist them. A variety of methods can be used to keep the motoring public safe without taking advantage of the situation. It is recommended that officers work closely with highway engineers to determine their schedule for any studies and reposting of the highway. Also high-visibility patrols can be very effective. Remember in Chapter 2 we talked about the fact that most motorists do not know why a car is stopped and, as such, any stop can be a deterrent to speeding. Finally, a shift in policy to the road situation can keep the motorist cooperative in obeying the speed laws. A traffic stop that results in a seat belt violation or other nonmoving violation can be effective in getting everyone to slow down until the correct signs are in place. It is important to note that when a new limit is posted, a period of time should be designated to allow for warnings to be issued. In this way, motorists can get used to the new limit.

The last cause is a result of the 85th percentile. As you recall the process, consider a road that has a volume of 100,000 vehicles a day. If the speed limit is established based on the speed 85 percent of the motorists are comfortable with, then those going faster are the motorists who need to be reminded to slow down. In the situation with 100,000 vehicles a day, 15 percent would be 15,000 vehicles that will be going over the limit. Since that includes those going at least 1 mph

(1.6 km/h) over the limit, the officers would apply their own discretion. This results in at least 7,500 vehicles traveling at a speed which can result in a citation. There is not an agency in existence that has sufficient staffing to handle that many speeders. Adjustments in tactics should be made to try and get voluntary compliance. Educating the public has an important place here. In 1974, when the National Speed Law went into effect, state troopers across the country had to drastically increase their speed enforcement efforts to reduce the speed of approximately 50 percent of all motorists using the interstate system. In Florida, the highway patrol formed two groups of troopers called the "dirty dozen." Their procedure was to stop as many motorists as possible who were violating the speed law. Federal guidelines dictated that citations had to be written, but with 24 troopers working over 2,000 miles of roadway while still handling their regular duties, it was an ominous task. Utilizing a "hit and run" approach of working a stretch of interstate, disappearing, and reappearing on a new stretch of road allowed them to be viewed by many motorists. By the end of the first two weeks of this campaign, it was nearly impossible to find a speeding vehicle.

Chapter 4

BASICS OF DETERMINING SPEED

Once a speed limit has been set, the law enforcement officer needs to keep an eye open for those who would drive in excess of the posted speed limits. But how does the officer make a determination that a vehicle has exceeded the speed limit? Can an officer simply observe a vehicle and make an assertion that the vehicle has traveled in excess of the posted limit? These are interesting questions and even though I would answer them with a "yes," the officer needs to convince more than me. The driver of the vehicle needs to be convinced of the allegation, and the courts need to agree with the officer. This means that the officer must present some sort of evidence to the fact that the vehicle in question was speeding.

THE PACE CLOCK

A pace clock, often referred to as a speedometer clock, is made as the law enforcement vehicle travels at a speed very close to the speed of another car which is the target of the officer's attention. Traditionally, the target vehicle is directly in front of the officer; however, the same type of action can be taken if the target vehicle is behind or even beside the officer's vehicle. Notice I used the term *very close* to the speed of the target vehicle, as it is very difficult to match the speed exactly. Actually there are three stages in a pace clock:

- Stage 1–First the officer must suspect that the target vehicle is in violation of the legal speed limit. Remember that at the beginning of this section, I said that I believe an officer can tell that a vehicle is in fact speeding. Part of this comes from the learned behavior the officer gains from driving around on patrol, and part is from the practice of estimating the speed of vehicles, which is done very often as the officer patrols. This is not difficult since the officer should be driving at or below the posted limit while on patrol. While driving at the speed limit, the officer is watching for vehicles that appear to be traveling faster than the speed limit.

- Stage 2–Next the officer will accelerate to reach the speed of the target vehicle. This method of catching up to the target vehicle has a degree of danger attached to it. The officer needs to be aware of the other traffic and potential road hazards while not losing sight of the target vehicle. It is acknowledged that the law enforcement officer most often has better training and skills at driving than the civilian driver. But this stage is very critical. This is very close to a chase and there are plenty of civil litigations involving police chases. The officer must constantly refresh the assessment of the situation and be prepared to discontinue this course of action when the risk to innocent parties becomes too great.
- Stage 3–Finally the officer must set the speed of the law enforcement vehicle to the same speed as the target vehicle. This is not a momentary event, but the matched speeds should exist long enough to establish the target vehicle's speed. Often, courts will expect a certain distance such as a block or a certain time period such as five to 10 seconds. If this is what the courts want, then every effort should be made to comply with their wishes, but the time period simply needs to be long enough to establish the speed. Keep in mind that too long and the present danger of a target vehicle driving too fast could end in a crash. Finally, the determination may be a relative assessment. As I stated, an exact match is difficult. The number of mph (km/h) difference between the law enforcement vehicle and the target vehicle will vary with each given situation. The sensing of this action is relative to the ability of the officer to judge movement and the environment this pace clock is being conducted in. Some officers can judge speed to within only a few mph (km/h), while others may not be able to do as well. It should be understood that the environment plays a role in the pace clock. Darkness and narrow

streets can have the officer overestimating speeds. It is recommended that in a pace clock, once the officer reaches what is believed to be the matched speed, then the officer should reduce the speed of the law enforcement vehicle, allowing the target vehicle to start to pull ahead slightly. Using the speed determined at that point, the officer can say that the target vehicle was traveling "at least" that speed.

The speedometer in question on the law enforcement vehicle needs to be accurate. Yes, they do come with the word "Certified" on them, but they can give erroneous information. The speedometer needs to be checked by a speedometer test shop on initial use and at least every six months thereafter. And this speedometer shop must be capable of repairing a speedometer if it is found out of tolerances.

STOPWATCH METHODS

In Chapter 2, we discussed the definition of speed to be distance divided by time. It is therefore natural to use a method of computing speed from known time and distance. If a vehicle passes point "A" and later point "B" and we know the distance between these two points and the time it takes to travel that distance, we can determine the average speed. The important part of this method is to take accurate measurements of the distance and the time. Some of the earliest methods were to use pneumatic hoses across the road. The hoses are placed a certain distance apart and then when the vehicle drives over the hoses it starts and stops a timer. The information is calculated and the resulting speed is determined. The biggest disadvantage to this method is that the set up time is excessive and the device cannot be left unattended for fear of theft or vandalism.

The measurement of distance was generally done with a tape measure when the distances were 300 feet (100 meters) or less. But when the distances were over 300 feet, a vehicle was generally used. Of course, there are some inherent problems with this method of measuring. Odometers are not always accurate and are subject to tires on the vehicle. The distance is very dependent on the ability to see the vehicle cross both reference points. Often an officer would find two

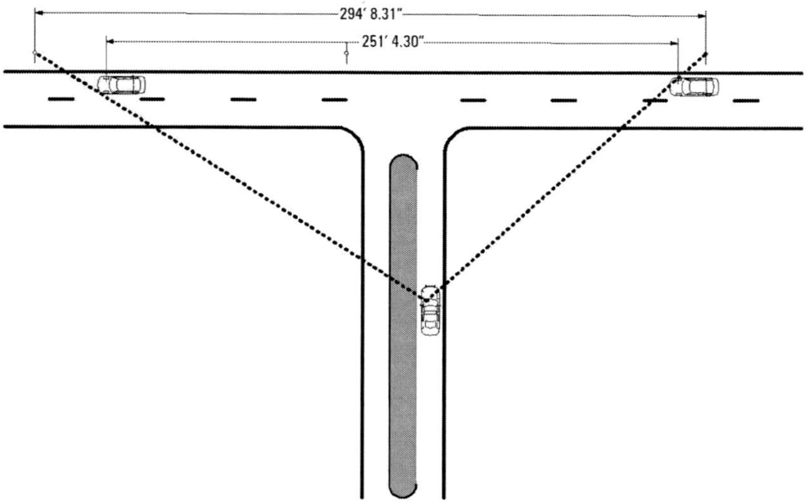

Figure 5. Parallax error.

poles that could be used as reference points. The officer would position his vehicle on the opposite side of the road between these two points and as a target vehicle would pass them, a timing device would be turned on and off recording the time it took to drive between these points. Early on a serious problem with the way of measuring speed was found. When an officer uses points the vehicle passes but does not touch, an effect known as parallax would cause the readings to be off. Figure 5 shows the error inherent with the process.

Parallax occurs when two objects are not lined up properly to the eye of the person viewing these objects. Sharpshooters are aware of this effect and compensate for it when sighting a rifle. It can be most visible in a car. The driver sees the speed on the speedometer accurately because he/she is directly in line with the speedometer needle and speed indicators. But a passenger will see the speedometer reading low (in the United States[1]) because he/she is situated to the right of the speedometer.

To correct for parallax, the points the target vehicle passes should actually touch the vehicle. For example, lines across the roadway or changes in pavement textures are best to use as reference points.

1. In countries where the vehicle is operated from the opposite side and the passenger is on the left, they will see the speed as higher than actual.

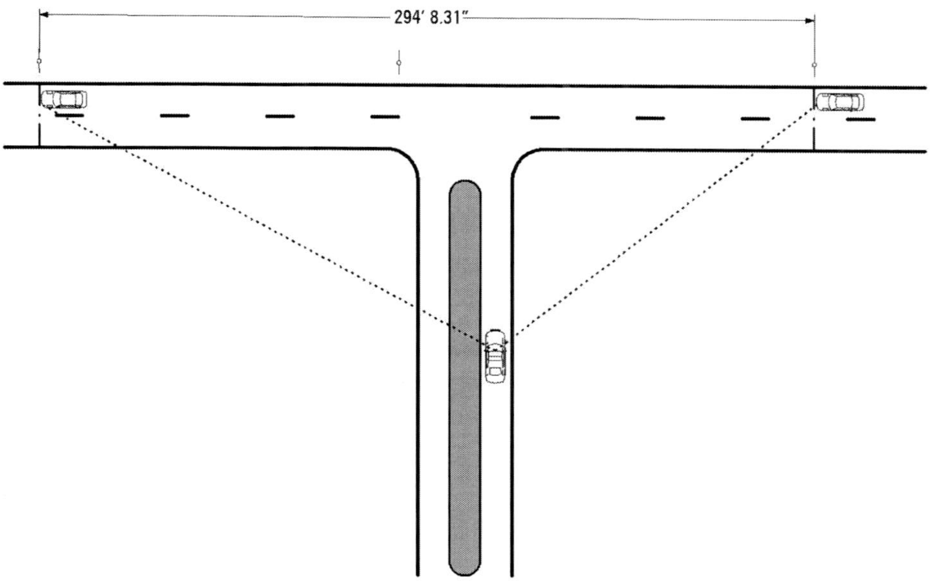

Figure 6. Parallax is eliminated through proper reference points.

Whatever point is selected, it needs to be a fixed point. Shadows, for example, change position with the time of day. In Figure 6, you can see that the vehicle is timed across the exact distance measures and there is no error in the calculation.

Also, the distance must be long enough to allow for human error in timing. I spoke with an officer once who used the width of an intersection to time vehicles for speed enforcement in a 30 mph (48.28 km/h) zone. White lines had been placed on either side of the intersection at a distance of 44 feet (13.41 meters) apart. The officer's rationale was that it would take one second for a vehicle to pass between the lines; anything less would be speeding. The problem was that the potential for human error in timing would be very likely. Unless the officer was exact on the timing, an error of one-half second could actually double the speed. It is recommended that when using a distance to time a vehicle and the timing device is being activated manually, the distance should not be less than 150 feet (45.7 meters) and it is preferred to use 300 feet (91.4 meters) or more. And, follow the rule that the higher the anticipated speed, the longer the distance should be. A vehicle traveling 60 mph (96 km/h) over a distance of 300 feet (100 meters) will take almost 3.5 seconds to travel that distance. It is reasonable to assume that the operator of the manual timing device would see if the vehicle was missed by one-half of a car length. If a car length is 16 feet (5 meters), a one-half car length miss early would raise the target vehicle speed 2.5 mph (4 km/h). Double that distance and the error time would be about 1.25 mph (2 km/h).

The other half of this method is the stopwatch. It too must be accurate, but the accuracy is much easier to validate and much more dependable. Most stopwatches are accurate to within one-half of 1 percent. That means that the error factor at its greatest is 5/1000th of a second. The problem is not as much the device as it is the human operating it. But this is overcome by a thing called anticipation. Like a runner being timed on a track, the human eye watches the approach to the finish line and anticipates the runner crossing that line. This makes the timing very accurate. Some agencies require that the person operating the stopwatch allow a car length to pass before ending the time. By doing so, both distance and time error potentials are compensated for. I do not endorse this means of error compensating as I feel the officer can get very precise information for the calculations needed, and the potential error is eliminated by the officer's integrity in not attempting to prosecute someone who is not clearly in violation of the speed laws.

Once we have the distance and time information, what do we do with it? What is required is for the information to be calculated to a speed. This is done by using the formula we learned in Chapter 2:

U.S. (Distance Traveled) / (Time Elapsed) / (5280/3600) = Speed in mph

Metric (Distance Traveled) / (Time Elapsed) / (1000/3600) = Speed in km/h

Of course, it would not be effective to sit on the side of the road and calculate each speed as the vehicles pass. Rather, it is best to have a chart of speeds for a given distance (see Table 3). Then it is a simple matter to look for a vehicle that goes across the distance in less time than the allowable limit. If the speed limit is 35 mph (55 km/h) and the desired speed to cite is 50 mph (80 km/h), then any vehicle going through our distance in less time than 4.1 seconds (for a distance of 300 feet or 4.5 seconds for 100 meters) would be at a violation speed. The chart can be used after the stop to assign a speed to the violation.

Table 3.
TIME DISTANCE SPEED CALCULATIONS

SPEED CHART–U.S.			*SPEED CHART–METRIC*		
DISTANCE (IN FEET)	*TIME*	*SPEED (MPH)*	*DISTANCE (METERS)*	*TIME*	*SPEED (KM/)*
300	4.5	45.5	100	4.5	80
300	4.4	46.5	100	4.4	81.8
300	4.3	47.6	100	4.3	83.7
300	4.2	48.7	100	4.2	85.7
300	4.1	49.9	100	4.1	76.8
300	4.0	51.1	100	4.0	90.0
300	3.9	52.4	100	3.9	92.3
300	3.8	53.8	100	3.8	94.7
300	3.7	55.3	100	3.7	97.3
300	3.6	56.8	100	3.6	100.0
300	3.5	58.4	100	3.5	102.9
300	3.4	60.2	100	3.4	105.9
300	3.3	62.0	100	3.3	109.1
300	3.2	63.9	100	3.2	112.5
300	3.1	66.0	100	3.1	116.1
300	3.0	68.2	100	3.0	120.0

Let's try an example; If I only wish to stop vehicles for 50 mph (or 80 km/h) then I only need to look for a time that a target vehicle travels through my distance at those speeds or faster. In the U.S. table, the time would be 4.0 seconds and in the metric table, the time would be 4.5. So if a target vehicle traveled between the two points in 3.6 seconds, I would not need to do any calculations. I already know the vehicle is speeding. Once the vehicle is stopped, I can look up the speed based on the time for the citation.

EARLY TIME DISTANCE CALCULATORS

The TDS (Time–Distance–Speed) unit from Smith and Wesson pictured was one of the first attempts to build a device that could record distances with a law enforcement vehicle and then time a selected target vehicle over that distance. One of the features of this device was that the distance could be locked in and reused over and over again. In this way, the officer could record the distance and then chase

Figure 7. An early time-distance speed computer.

the target vehicle, issue a citation, and then return to the same spot without having to rerecord the distance again.

Vascar by Traffic Safety Systems is an upgrade of the TDS system. Vascar is an acronym for Visual Averaging Speed Computer and Recorder. There were several models of these units made: Vascar, Vascar Plus, and Vascar III. These units were mounted in a law enforcement vehicle, attached to the speedometer cable, and used to record times and distances receiving digital speed readout on the display screen. In addition, the recorded time and distance could be recalled to the display for reference. The Vascar operator's manual indicates that there are five basic clocking methods used by the operator of a Vascar system:

Following: This is a traditional method where the law enforcement vehicle is following the target vehicle.

Opposite Direction: The target vehicle is approaching the law enforcement vehicle from the opposite direction.

Parked: The operator has already recorded a distance and then stopped on the side of the road waiting for a passing target vehicle.

"T" Intersection: The law enforcement vehicle is parked perpendicular to the road and as the target vehicle passes, the operator pulls out to establish a following clock.

Dial-A-Distance: The operator records a distance with the unit, recalls that distance, and programs it into the Vascar. Then as vehicles pass the reference points, they are timed. The advantage to this is that the distance remains in the unit memory until the distance switches are activated again, allowing the operator to clock a vehicle, pursue and stop that vehicle, and then return to the same observation point for the next clock. Also, if the operator writes down the distance, then the unit can be programmed again without driving that distance, allowing the operator to leave and return later to continue clocking vehicles.

The Vascar unit was mounted in the law enforcement vehicle. The unit was attached to the vehicle's speedometer cable. Inside the cable assembly was a small disk with holes in it. On one side of the disk was a light, and on the other side was a photo cell. As the vehicle was driven, the disk would rotate. This caused the light to flash at the photo cell or pulse. The unit would count these pulses electronically. The Vascar unit was programmed to regard each pulse as 1/10,000 of a mile, so 10,000 pulses would equal one mile. Understanding that there are inaccuracies in the vehicle method of measuring, the people at Traffic Safety Systems built into the Vascar unit a way to adjust for these inaccuracies. Using a series of steps, it is possible for the Vascar to determine the proper number of pulses for the vehicle it is mounted to. Using a measured quarter mile as a standard, the Vascar will count the appropriate pulses for an accurate mile and the operator simply inputs that information. For example, if the Vascar indicates that in a particular vehicle the correct number of pulses for a mile is 9,988 instead of 10,000, then the operator inputs that information to the unit's distance memory, and as long as the unit stays on, all calculations are adjusted to the correct distance. However, if you shut the unit off and turn it back on, the unit would revert to the 10,000 default count and the operator would have to recheck the unit by driving the measured quarter mile.

In addition, the operator needs to test the timing recorder in the unit for accuracy. In this case, the operator simply inputs a preset distance and then using the timing switch, records a period of time getting a speed. Checking this against a chart of distances, times, and speed is a simple matter of verifying the output information. This process should be conducted at least two times: before and after enforcement activity. Generally, this was at the beginning and ending of a tour of duty.[2] Operators of these units discovered that the number of pulses would change slightly from the before to the after tests. This was due to tires heating up and therefore a tolerance was required to be sure the differences were not too great. The Vascar operator manual indicated a tolerance of one-half of 1 percent to be acceptable; so if the starting pulse number was 9,988, the ending number would need to be between 9,983 and 9,993. If not, then every citation issued that day would have to be voided.

Notably, the distance portion of this calculation is the most difficult variable. This is not to say that a skilled operator could not determine accurate speeds. I used the Vascar Plus for many years and lost only one case in court. It was due to a distance portion of the calculation, but it was not an error in the distance per se. It was in the fact that I wrote down the wrong reference point in my notes. Otherwise, I found Vascar to be an effective tool for speed determination.

There have been other time distance calculators since Vascar. Radar and laser manufacturers have placed time-distance functions in their devices (we will discuss those in the appropriate sections) and in some cases have come out with stand-alone time distance calculators (i.e., the "Tracker" by Kustom Signals). One device that is used today is nothing more than a stopwatch that can calculate speed. The Robic Company manufactures a variety of sport stopwatches that are used for athletes and sporting events. It allows for input of a measured distance. Then when a time is collected by the watch, the resulting speed is displayed along with the time elapsed. The most attractive feature to this stopwatch is the price, which is less than $100 (possibly cheaper if bought on eBay). So you can see there are plenty of options. In fact, with hopes that I don't confuse you with this statement, you will find in the next two chapters that radars and lasers are, in essence, time distance calculators.

2. In Chapter 8, we will discuss the legal issues associated with testing.

Chapter 5

BASICS FOR POLICE TRAFFIC RADAR

Many of today's consumer products were not invented for the purpose they are used. Velcro®, invented for use in the space program, for example, can be found almost daily in all sorts of applications that have nothing to do with the space program. Likewise, radar, which was devised for military use, has a multitude of applications. We will review its history, and will progress through the speed detection applications physics and operational considerations.

RADAR'S HISTORICAL DEVELOPMENT

It is said that radar was more of a discovery than an invention. At the beginning of World War II, radio operators were in contact with their French counterparts across the English Channel. While in communication, a large passing ship would reflect back their signal, causing an "echo" of their transmissions. Feeling that this could be useful in detecting ships and planes, the British engineers started looking into this phenomenon. A group of engineers led by Sir Alexander Watson-Watt started working on the concept that they called RADAR. The word is an acronym for *RA*dio *D*etection *A*nd *R*anging and was descriptive of its function. Even the acronym showed the function of radar since the word is a palindrome, a word which is spelled the same forward or backward. The British-built radar transmitters and receivers along the coast to detect aircraft and report their position, speed, and direction. Twenty-one towers were erected and dubbed the "Chain Home Network." These towers could detect an aircraft from about 100 miles away. Upon gathering this information, it would be relayed to the command center, which would alert British fighters to respond and intercept the aircraft before they reached their target.

In the summer of 1940, the Battle of Britain took place. The odds were against the English due to the number of planes, but the early detection of the German aircraft allowed the British to fly out and surprise the enemy. By the end of this siege Hitler, who was seeking air superiority in order to support an invasion of England, backed away from his plans. In a statement concerning the value radar gave to the Royal Air Force, Wing Commander NWP Hancock of the Royal Air Force recounted, "If the air war had been lost and the Germans had come here and had been successful; a lot of if's. But if it had happened, well then I wouldn't be here and you would be speaking German."

During the ensuing four years from 1940 to 1944, radar development accelerated. Taffy Bowen brought the radar concept, and a new component, to the United States over a year before the U.S. entered the war. Working with engineers from the Massachusetts Institute of Technology, Mr. Bowen introduced them to the "Cavity Magnetron" which was able to produce radar waves at a much shorter wave length called *microwaves.* This method allowed the radars to be built smaller, giving them the potential to be mounted on aircraft and ships. One such radar unit was mounted on the British Aircraft Carrier HMS Ark Royal and in June of 1941, it was used to track the German Battleship Bismarck on a three-day chase through the Atlantic. The conditions of this chase were through stormy and foggy seas. Ultimately, the carrier Ark Royal was in close striking distance and on the third day, her aircraft were able to sink the Bismarck. Without the radar tracking system, the Bismarck would have undoubtedly escaped.

By the end of the war, engineers had developed 120 various types of radar designed to detect air and surface objects. After the war, further developments in radar allowed the ability to give even more information from returned radio waves. The original radar units were capable of giving the operator information on object speed, direction of travel, and distance. Such radars were used on December 7th, 1941 when the Japanese attacked Pearl Harbor. In that situation, the radars worked fine, but the operators and advice from superiors led to the misinterpretation of 353 Japanese zeros as returning U.S. Bombers. The radar worked properly in its capacity but was simply misunderstood as to what it was reporting. Radars now can determine size, shape, weight, and when coupled with computer technology, probable target identification. Dr. Norman Ramsey, professor of Physics at Harvard University, has stated that today's radars can detect an object 100 miles up and "determine if it is a Volkswagen or a Toyota."

After the war, radars met several civilian applications such as weather forecasting and air traffic control. Then, in the 1950s, an educator, Mr. Bryce Brown at Millikin University in Decatur, Illinois, developed a radar device for determining the speed of cars. This radar was an abbreviated device in as much as it could only determine speed. It required being operated from a fixed location when actively seeking speeds. Mr. Brown established his product through a Canadian company in 1955. Since then, many manufacturers have been born, merged, succeeded, and died. Each manufacturer has introduced law enforcement to variations of the radar: The MV-715, the K-55, the MR-7, and the ATR, which have given way to more dynamic names like the Genesis II, the Ranger, and the Raptor.

THE PHYSICS OF RADAR

In 1842, Austrian physicist Christian Johann Doppler (see Figure 8[1]) was interested in determining motion of stars. Through experiments that he conducted, he determined that transmitted waves changed their properties when they interacted with a moving object. Working with sounds, he concluded what is known today as the *Doppler Effect.* His experiments determined that waves such as sound waves increase in frequency when the listener is approaching the source of sound, and inversely, the waves decrease as the listener moves away from the source of the sound. This may be observed at any time when there is a sound source in motion or when you are in motion with relation to the sound source. For example, as a train approaches a railroad crossing, the engineer in the train will sound the horn warning of the approach. To the engineer, the sound is a steady tone. But to a motorist waiting at the crossing, the pitch is higher as the train approaches, and then the pitch is lower after the train passes. The amount of change in this pitch is dependent on how much speed is involved. Doppler had done this research for the determination of the movement of stars

1. Picture from University of Cambridge; author anonymous.

Figure 8. Christian Johann Doppler (1841).

which would relate to changes in star colors; something he could see. By determining the effect his principle had on sound, he could rationalize the same effect to light which also has wave properties. Later, when radar was developed in the 1940s engineers relied on this principle to develop the radars they needed at that time. Since our discussion centers on speed, we will examine this principle as it relates to speed detection (with a little discussion on ranging properties thrown in for seasoning).

Wave Theory

Radio waves use the same principles as sound and light waves. In this discussion, we will concentrate on waves transmitted from some electromagnetic source such as a radio or radar. As the word implies, the signal from our radar has a rise and fall to it, as well as a speed

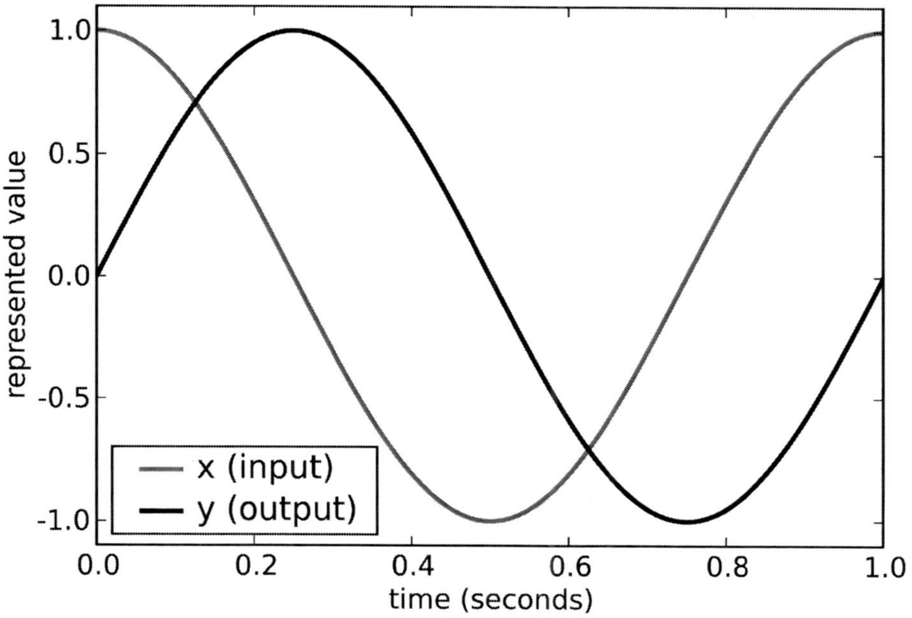

Figure 9. Diagram of a single wave. Note that the y (output) is the transmitted signal.

since it travels from and to the radar unit. Like the waves on the ocean, waves from electromagnetic sources have a cycle; a rise and then a fall, followed by a rise and then a fall, etc. A single wave may be described as a distance from peak to peak, or it may be described as a distance from the point at the beginning of the peak to a point at the end of the valley. Examining Figure 9, the *y* (output) signal shows the traditional view of a single wave. The *x* (input) signal is also considered a single wave, only described as peak-to-peak. In this manner you can see that a wave is cyclic.

A diagram of a wave sliced in half and then folded over would be a circle. With this in mind, we need to consider the three properties that define a wave:

The first consideration is that all waves generated from an electromagnetic source travel at the speed of light. In high school, you may have learned about the speed of light. The speed of light is a constant which has been mathematically calculated to be 186,282 miles per second (299,792.458 kilometers per second). The wave transmitted from a radio or radar unit is therefore traveling at the speed of light, and for the sake of this discussion, that speed does not change.

The next consideration is that electromagnetic waves have a frequency. If you look at the root word of frequency, which is frequent, then you will recognize that the word indicates "how much." Remember, we described this signal as a wave, which is to say that there is a rise and fall to characterize what the signal is doing. An electromagnetic signal will have a certain number of these waves in an interval of time. The number of those waves is referred to as its frequency. The interval of time is universally set at one second, so when we speak of a signal as having a frequency we are saying it has a certain number of waves each second. A frequency is one way to describe an electromagnetic signal. For example, when you listen to a radio station, you will tune your radio to the frequency for that station. If you are listing to a station at 840 AM, you are listening to a station that is transmitting at 840 kilohertz, or 840,000 waves per second. Likewise if you are listening to 106.5 FM, you are listening to 106.5 megahertz, or 106,500,000 waves per second.

Finally, all waves have a length. Since electromagnetic waves have speed, they will end up traveling a distance over time. If we transmit one wave, then in one second, that wave will have a length of 186,282 miles (299,792.458 km). The waves in Figure 9 represent this distance as shown. If we transmit two waves in one second, each wave will have a length of 93,141 miles (149,896.229 km). Notice the result of two waves when added together equal the distance light traveled in one second.

Next, we will take what we have discussed in these three properties of waves and make some observations.

Let us understand a few synonymous terms that we will be using. We have said that a signal from an electromagnetic source has a certain number of waves in a given second. We have also said that a wave is a cycle. Therefore, we can describe this signal's frequency as a number of waves per second, or a number of cycles per second. Still, both of these terms are somewhat lengthy, but thanks to physicist Heinrich Hertz we can shorten it up to just Hertz. So our wave can be described as:

Waves per Second (i.e., 50 waves per second), or
Cycles per Second (i.e., 50 cycles per seconds), or
Hertz (i.e., 50 Hertz)

Also, these terms can easily become lengthy. In our FM Radio example, the designated frequency was 106.5 FM. Appling this to our expressions of waves, cycles, and hertz, this frequency may be expressed as:

106,500,000 waves per second, or
106,500,000 cycles per second, or
106,500,000 Hertz

But these are still a bit lengthy, so we can use metric shorthand to make them easier to express. This metric shorthand is an application of certain terms that describe quantity. They are:

Kilo (K) which means thousand (1,000)
Mega (M) which means million (1,000,000) and
Giga (G) which means billion (1,000,000,000)

You may recognize these terms, especially if you use a computer. Using this shorthand, our radio frequency is commonly expressed in megahertz, or as:

106.5 Megahertz

Finally, the terms have abbreviations as well, which are shown in our examples. In the case of the radio station the frequency is then described as 106.5 MHz.[2]

Now, understand that there is an inverse relationship between the properties involving frequency and wave length. If we increase the frequency of waves from one to two, the length of each wave must decrease in order to fit into one second. Therefore, the relationship is that the higher the frequency, the shorter the wave length. Remember that all waves are measured on the basis of one second. So a signal that

2. An uppercase H and a lower case z are used to signify the term Hertz, i.e., Hz.

has a frequency of 50 waves in one second has shorter wave lengths than a signal that has 10 waves in one second. Remember our radio at 106.5 MHz? That frequency sets each wave to a length of approximately 9 feet 3 inches (2.82 meters). This inverse relationship can be expressed mathematically as:

Frequency times the Wave Length equals the Speed of Light, or
F x WL = C
(Where F = Frequency, WL = wave length, and C = Speed of Light)

In Table 4, we see examples of frequency and wave lengths for typical common bands of transmitters, but there is another property of radio waves that can be investigated now that we see their differences. Have you ever asked yourself, how does the radio signal go through solid objects? I was investigating a bank alarm call and had an occasion to go into the vault. While standing in the vault, my dispatcher called me on the radio. We take it for granted that it works, but why does it work? The radio signal cannot bend or curve. Based in the laws of physics, it must travel in a straight line. It must then have the ability to penetrate objects. Understanding this property will help you understand how your radar works.

Table 4.
COMPARISONS OF COMMON FREQUENCIES AND WAVELENGTHS

ITEM	*FREQUENCY*	*WAVE LENGTH (U.S.)*	*WAVE LENGTH (METRIC)*
AM RADIO (TYPICAL)	690 KHz	1,425 FEET	434.34 METERS
FM RADIO (TYPICAL)	106.5 MHz	9.25 FEET	2.82 METERS
POLICE VHF RADIO	166 MHz	5.93 FEET	1.81 METERS
POLICE UHF RADIO	460 MHz	2.414 FEET	.652 METERS
POLICE 800 SERIES RADIO	800 MHz	1.23 FEET	.375 METERS

A radio wave will travel forever. In a vacuum, the radio will sustain the speed of light forever. However, things can interfere or change a radio wave. For our discussion on radio waves as they relate to radar, we will look at three things that can change a radio wave's travel.

1. Reflection–This occurs when a radio wave's direction is changed and folded back towards the source. Sound waves reflect when you shout into a canyon, for example. It appears that you can hear your own voice returning to you. Actually these are reflected sound waves. Different substances will return the signal better or worse. For example, aluminum and steel return the signal better than fiberglass or wood. Also, polished surfaces give a better return signal than dull or dirty ones.

2. Refraction–This is similar to reflection, but the signal is not returned. Instead, it is simply bent to a new course of travel. When radar passes through glass, for example, there may be a degree of refraction, or bending in the direction of the wave. This is due to the change in the medium that the wave is passing through. As a radar signal passes through a windshield from a dash-mounted antenna, the signal may experience a slight refraction, but not enough to give false information. But what if the glass is flawed? The best determination of refraction through glass is to look through it when you place the antenna. If the images of things on the other side are distorted, then the radar signal will be distorted.

3. Absorbed–A radio signal will travel forever unless it is absorbed. Many substances can have different absorption properties on a radio wave. To an extent, the radar signal is absorbed by water and water vapor as well as plant life. Absorption might be slight, or it might be extreme depending on the object and the proximity of the radar to them. Most objects that have the tendency to absorb will also reflect some of the energy.

In our types of radio transmitters, we have different range properties. Radios transmitting in the AM bands have greater range than the ones in FM bands. Under certain atmospheric conditions, an AM radio can be received within several states, while FM radios can only cover several counties at best. Likewise, police VHF radios have greater range than FM or 800 Series radios. The ability of the longer wavelength radio transmitters to penetrate great distances is readily seen. HAM radio operators use wavelengths all the time to reach great distances. By "bouncing" signals off the atmosphere, they can pass line of sight distances and reach around the world. Obviously one of the more desired qualities of radar is the ability to get a clear echo. It stands to reason that the higher the frequency, the better the reflection will be.

One thing I would like to touch on is the concept of amplitude. Amplitude is related to the power of the transmitter. In our wave, increasing the power would increase the height and depth of the wave from the centerline. To put it more simplistically, if we increase the power of a wave signal, we increase its strength and thereby its energy. Radars for traffic enforcement do not require high power and most are manufactured at an output of less than five miliwatts per centimeter squared (written as 5mw/cm2). Compare 5mw to a microwave oven that may operate at 100 watts or 2,000 times more power than police traffic radar. Recognize there are more powerful radars. Radars are made to track aircraft at high altitudes and even satellites in orbit, and such radars have tremendous power-output levels.

The other property we need to understand when comparing frequencies and wavelengths is that as we increase frequency, we also increase the directional properties of the signal. At lower frequencies, radio signals trend to spread more rapidly than at higher ones. Understand that this is not a naturally dramatic property, but with proper antenna applications it can be very beneficial to radar use. Recall that Taffy Bowen brought a device to M.I.T. called a cavity magnetron, a device designed to produce microwaves. As the name implies, microwaves are very short waves. The definition of a microwave frequency is complex and is identified by the Federal Communications Commission (FCC). For our purposes, we will only discuss microwaves as they relate to police radar. To the FCC, the correct term is *radio locating transmitters,* and they assign the frequencies in which their transmitters will operate. In viewing Table 5, you can see the various frequencies assigned to radio location devices for civilian use. Those listing in bold are the most common police radar frequencies in the United States, although the others may be used in other countries.

You can see that the frequency of radar for traffic law enforcement is very high. The designations provided by the FCC are not arbitrary. They are set in response to the radar manufacturers themselves. You will take note of the Ka Band radars for example. It is the only radar

Table 5.
RADIO AND MICROWAVE FREQUENCY BANDS FOR POLICE TRAFFIC RADAR[3]

BAND	*FREQUENCY*	*WAVELENGTH*	*NOTES*
S	2.455 GHz	4.8 inches (12 cm)	Obsolete
X	**10.525 GHz**	**1.1 inches (2.8 cm)**	**Current production of these radars is limited.**
Ku	13.450 GHz	0.88 inches (2.2 cm)	No known systems
K	**24.150 GHz**	**0.49 inches (1.2 cm)**	
Ka	**33.4 to 36.0 GHz**	**0.35–0.33 inches (9–8.3 mm)**	

frequency that is set to a band. The concept at the time was to allow the manufacturers the ability to change the frequency in different model radars so as to prevent the radar detector industry from manufacturing effective radar detectors. Today, a variety of radars operate on different frequencies in that band. All major U.S. manufacturers of police traffic radar use the Ka frequency bands. In addition most photo radar companies operate in the Ka band of frequencies.

The final property that we need to be aware of is that the beams of radiated signals have width. Like range, width is regulated in part by the antenna, and in part by the operational frequency. There are no police traffic radars with a transmitted beam so narrow that a single vehicle can be picked out of a group of cars on the highway. But the beams are narrower with higher frequency radars. I have been involved in a multitude of training programs where time is spent on identification of the width of a radar beam and I find this information fleeting to the student. Take a standard highway with 12-foot (3.65 meters) wide lanes. A two-lane road is then 24 feet (7.3 meters) wide, a four-lane road with a 15-foot safety zone is 63 feet (19.2 meters) wide, and an interstate with a 60 foot (18.3 meter) median is 108 feet (31.2 meters) wide. Regardless of the radar used, the entire road is bathed in the radar beam at 1,000 (304.8 meters) feet. For those who need to know, I will give you a ballpark idea of how wide these radar beams

3. Source: Federal Communications Commission, December, 2001.

are. Most often, a target will become apparent at a distance of around 100 feet (304.8 meters). At that distance, the three most common radar frequencies will be the following width:

X-Band will be 314 feet (95 meters) wide
K-Band will be 208 feet (63.4 meters) wide
Ka-Band will be 158 feet (48.2 meters) wide

The strongest part of the beam is toward the center and becomes slightly weaker as you move towards the edges. Vehicles in the main portion of the beam give a better returned signal than those in the outer edges of the beam.

Radar Decision Making

The radar signal's characteristics we have been discussing establish its ability to display a speed. But which speed? Now that we have an understanding of these characteristics, we can turn to the things that make a vehicle a good (or bad) target for the radar. There are three factors that set up the radar's decision-making ability. Understanding these three considerations will help the operator determine which vehicle is related to the speed display on the radar. These properties are:

1. Reflective Capability–For some reason, this is one of the more difficult concepts for some to understand. Reflective capability covers all aspects of the target vehicle that causes it to be more or less reflective. Included in the determination of the degree of reflectivity are surface area exposed to the radar signal, the shape (aerodynamics) of the target, and its composition (how dense is the material it is made from). In a small way, polish has a role; but for radar, not a very large role. People think if they leave their vehicle dirty, the radar will not see them. This is false thinking. Remember that radar waves are very small. Dirt particles and dullness in the paint can make a vehicle reflect more of a radar signal.

2. Position–It is not a mistake that Reflective Capability is the first consideration; for without it being understood, there could be no comprehension of the factor we call position. It is not just how close the target vehicle is to the radar, but it is also how much energy it is returning. A vehicle that has 25 square feet (7.62 square meters) of surface

area reflects back only that radiated energy that hits it. If that vehicle is so close to the radar as to reflect back 80 percent of the signal, the energy returned will be significant and that vehicle will be displayed over other targets. But at a 1,000 foot (304.8 meter) range, the reflected energy may be only 15 percent which would be a very weak return. You may have seen this in a motorcycle attempting to overtake a large truck. The radar may never display the motorcycle as the returned radiated energy does not exceed that of the truck, which is closer and larger than the motorcycle. Radar with a beam that is not as wide will allow a larger percentage of energy returned for the same distance.

3. Speed–This final factor is not as prominent as the other two, but it does have its place. When all other factors are equal, speed becomes an important element in the determination of which target to display. The radar is looking for the strongest return signal and if all the return signals are about the same, a fast-moving target will be sending more energy back to the radar. A target vehicle moving at 100 mph (160 km/h) is sending a strong return signal. The radar will often display that vehicle even if it is farther away that the others depending on the other vehicles' reflective capability.

It is important to keep in mind that we are extracting these properties which are important to the physics of radar, but there are other properties that we have not addressed. These other properties are not as important, but they do have an impact on the way radar functions on earth. If we represent the edge of a radar beam by two lines at angles to the radar, understand that the actual edges do not look like two straight lines; rather, they are erratic and in constant flux due to the other properties we have not discussed here. This section is presented to bring your understanding of radar to a particular level without your becoming a physicist in the process.

Precision, Accuracy, and Reliability

Often the question may be asked, "How reliable is the radar?" This is a common question because if we are determining speed in order to accuse someone of going in excess of a legal limit, we need to know that the information provided by radar is "fair and accurate" in its introduction as evidence. So let's look at radar accuracy.

If you enter the number 8 into a calculator, and then divide that number by 2, the calculator responds with the number 4. That would be an accurate response. Any other response would be *inaccurate.* Therefore the result must be free from error. The calculator cannot be *almost* correct. Radars are like the calculator in that the response is either correct or incorrect. But radars are dependent on input information. Accuracy is only maintained if the results can be explained based upon the input variables. This can only be verified by the testimony of witnesses (the operator). Things such as angles between the radar and the target vehicle, or outside radio signals can affect the displayed reading. However, the reading is still an accurate result of the information given to the radar. The result has been adjusted by some outside factor. The radar must perform according to physical laws. The radar operator must have sufficient training and experience to be able to discern the source of the radar's displayed reading (target vehicle or not) and if that reading has been adjusted by some factor (i.e., cosine).

Based on what we have stated so far, it can be seen that radar's accuracy is not expressed as a measure or degree of accuracy; it is simply accurate. But there is a measure called precision. Precision applies to the result, and not of the device. For example, we enter the number 7 in the calculator and then ask for the square root. The result is 2.645751311. But maybe our calculator is set to only two decimal places. Then the result would be 2.65 instead (in this instance we rounded the answer based on certain rules. Those rules state that a decimal number is rounded *down* for a decimal of 1 through 4, and rounded *up* for decimals of 5 through 9). Both answers are accurate, but how precise? If we square the result, we should get 7 again. But squaring 2.65 gives us 7.02, which is off by .02 from the exact answer. The result of 2.65 may be precise enough depending on what we are measuring. If we are launching the space shuttle, for example, we may wish to extend the answer the full nine decimal places.

Engineers who designed the radar for law enforcement elected to work in whole numbers, numbers without decimal results. If we applied the rounding rules, then a number like 43.5 must be rounded up to 44, which certainly is not a fair and accurate represented speed of the vehicle. In radar, the decimal is truncated or simply removed. Radar's precision is a whole number value. Thus, a radar speed reading will always be within 9/10th of the exact speed, and will never display higher that the exact speed of the target vehicle. This applies to all radar displayed speed measurements for law enforcement application. The radar will be within 9/10th of a mile per hour; within 9/10th of a kilometer per hour; and even within 9/10th of a nautical mile per hour. The system of measure is programmed into the radar, but the math stays the same. Note that there are radars that do give decimal speeds. They are used for things like water flow assessments and other scientific, nonlaw-enforcement applications.

The final question may be how reliable is the radar speed reading. Reliability is the extent to which an experiment, test, or measuring procedure yields the same results on repeated trials. In our calculator, we can enter an operation 100,000 times and get the same answer. The same is true for radar. Radar's reliability is the result of certain testing which will be discussed later. In the following material, any examples which result in a final speed will be given in whole numbers that have been truncated.

RADAR COMPUTATION METHODS

The Doppler Shift

As previously stated, police traffic radar detects motion, and it reports the detection of motion as speed. This is done by applying the Doppler Effect. Recall that Doppler discovered if a source of transmitted energy and an object are moving closer together, the frequency from the transmitted source will increase. Assuming that the radar is in a fixed location (stationary) and the object moves towards a transmitted signal, such a movement will result in an increased reflected signal. The amount of change experienced is referred to as the *Doppler Shift.* The rate of change is uniform for each mile per hour (kilometers per hour) depending on the frequency of the radar used. In a U.S. system of measure, a speed of 1 mph will result in a:

Doppler Shift of 31.4 hertz in radar operating in X band (10.525 GHz)

Doppler Shift of 72 hertz in radar operating in K Band (24.150 GHz)

Doppler Shift of between 99.6 and 107.4 hertz in radar operating in Ka Band (33.4 to 36 GHz)

In a metric system of measure, a speed of 1 km/h will result in a:

Doppler Shift of 19.5 hertz in radar operating in X band (10.525 GHz)
Doppler Shift of 44.75 hertz in radar operating in K Band (24.150
Doppler Shift of between 61.9 and 66.75 hertz in radar operating in Ka Band (33.4 to 36 GHz)

So, let's look at an example of the process the radar follows to determine a speed. Because of the differences in the Doppler Shifts from U.S. to metric, we will use separate examples.

Example of Speed Determination–
Vehicle Approaching (U.S. Calculations)

A vehicle approaches the radar at a speed of 50 mph. The radar is transmitting in the K-Band at 24,150,000,000 hertz. The transmitted signal is reflected from the vehicle and undergoes a shift based on the direction of travel and speed. The signal is increased by 3,600 hertz, and returns as 24,150,003,600 hertz. The radar computes the difference between the transmitted and received frequencies and then divides them by the standard for 1 mph. The process would be mathematically expressed as:

Step 1–24,150,003,600 - 24,150,000,000 = 3600
Step 2–3600 / 72 = 50
Step 3–Display 50 mph as a target speed

Example of Speed Determination–
Vehicle Approaching (Metric Calculations)

A vehicle approaches the radar at a speed of 80 km/h. The radar is transmitting in the K-Band at 24,150,000,000 hertz. The transmitted signal is reflected from the vehicle and undergoes a shift based on the direction of travel and speed. The signal is increased by 3,580 hertz, and returns as 24,150,003,580 hertz. The radar computes the difference between the transmitted and received frequencies and then divides them by the standard for 1 mph. The process would be mathematically expressed as:

Step 1–24,150,003,580 - 24,150,000,000 = 3580
Step 2–3580 / 44.75 = 80
Step 3–Display 80 km/h as a target speed

The question now is what happens when the target is moving away? In this case, the radar sees a moving target and the reflected frequency is lower than the transmitted frequency. Again, the radar computes the difference and determines the speed. It is important here to know that the radar does not use negative numbers. The transmitted frequencies are quantities as opposed to values so the radar simply determines the difference. Taking our example, we will change the situation to a vehicle moving away from the radar. To help understand the new process, the changes in the previous example are italicized in this example. As before, we will use both a U.S. and a metric example.

Example of Speed Determination–Vehicle Receding (U.S. Calculations)

A vehicle is *driving away from* the radar at a speed of 50 mph. The radar is transmitting in the K-Band at 24,150,000,000 hertz. The transmitted signal is reflected from the vehicle and undergoes a shift based on the direction of travel and speed. The signal is *decreased* by 3,600 hertz, and returns as *24,149,996,400* hertz. The radar computes the difference between the transmitted and received frequencies and then divides them by the standard for 1 mph. The process would be mathematically expressed as:

Step 1–24,149,996,400 – 24,150,000,000 = | 3600 |
Step 2– | 3600 | = 3600
Step 3–3600 / 72 = 50
Step 4–Display 50 mph as a target speed

In the math formula above, you will notice the two vertical lines which represent the absolute value. This creates a new step where we change the absolute value to the numerical value that is treated as a positive number, and then we complete the process.

Example of Speed Determination–
Vehicle Receding (Metric Calculations)

A vehicle is *driving away from* the radar at a speed of 80 km/h. The radar is transmitting in the K-Band at 24,150,000,000 hertz. The transmitted signal is reflected from the vehicle and undergoes a shift based on the direction of travel and speed. The signal is decreased by 3,580 hertz, and returns as 24,149,996,420 hertz. The radar computes the difference between the transmitted and received frequencies and then divides them by the standard for 1 mph. The process would be mathematically expressed as:

Step 1–24,149,996,420 – 24,150,000,000 = | 3580 |
Step– | 3580 | = 3580
Step 3–3580 / 44.75 = 80
Step 4–Display 80 km/h as a target speed

In the math formula above, you will notice the two vertical lines which represent the absolute value. This creates a new step where we change the absolute value to the numerical value that is treated as a positive number, and then we complete the process.

Stationary Radar

Radar is made to detect motion. That would be its primary purpose. The inventors of the device used the acronym *RA*dio *D*etection *A*nd *R*anging as a way to describe its function. So radar detects and ranges. The motion it detects is referred to as relative motion; motion towards or away from the radar. If you are driving along the highway at a speed, your speed is relative to the other objects such as terrain and even the highway you are traveling over. If you were to hit a stationary object, the relative speed you were traveling would become very clear to you. But, if on this highway there is a car in front of you, traveling the exact same speed as you, there is ***no*** motion with respect to this other car. Radar would not see the other car since there is no relative motion. Also, the radar does not know if the motion it detects is the object or itself. The operator must either interpret what is happening, or give the radar more information. In our example of traveling along the highway, the radar will see the stationary terrain as movement relative to the radar. There are three basic situations we are concerned with in respect to relative motion:

1. Radar stationary–The object is moving towards or away from the radar.
2. Radar moving–The object is stationary.
3. Radar moving–The object is moving towards or away from the radar.

Radars designed not to move but detect an object's relative motion are stationary radars. Based on the way the radar has been manufactured, there is a minimum relative motion, or sensitivity needed so that radar can display a speed. Some manufacturers set the minimum allowed relative motion from as low as 2 mph (3 km/h) up to 15 mph (24 km/h). As we will find later in this book, this sensitivity can be adjusted by the operator in most radar units.

Radar speed is the result of motion towards or away from the radar, yet this motion may not be directly towards or away from the radar. This most often results in a speed that is lower than actual. We call this result the cosine effect. In physics, motion has two parts called vectors. With vectors, we can define how much motion is towards a particular point and how much lateral movement relates to that point. If a vehicle is driving 50 mph (80 km/h) directly toward the radar (not the most desirable situation to be in), then the direct speed is 50 (80) while the lateral speed is 0. Since we do not sit in line with the traffic, but rather along the side of the road, we must redefine the relative motion the radar sees. With our situation of a vehicle traveling at 50 mph (80 km/h), we now have the radar position 33 feet (10 meters) from the path of the vehicle (a lot safer). Now the relative motion is different than the actual motion of the target vehicle. The actual motion has not changed but the lateral motion is a negative quantity and will reduce the relative motion. The amount of change is dependent to the distances (how far down the road and how far off the road) involved. The relative speed in a situation where the radar is stationary is a result of applying the cosine of the actual speed and therefore is called the Stationary Cosine Effect.

If you have ever used radar and noticed that people seem to be slowing down as they approach you, understand that what you are

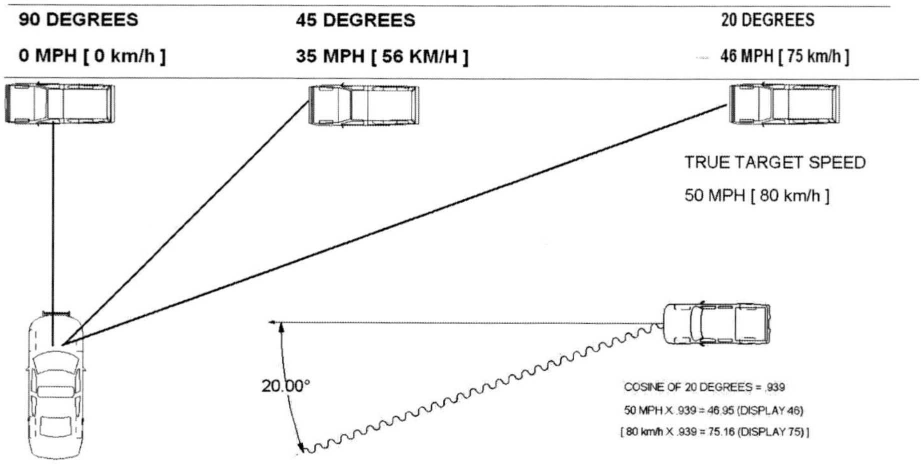

Figure 10. Stationary cosine effect.

observing is the relative speed of the vehicle to the radar due to the cosine of your positions (see Figure 10). It is important to recognize that in stationary operation, the effect from cosine is always to the advantage of the target vehicle. In normal stationary radar operation, the displayed speed will never be greater than the actual speed of the target.

Moving Radar

In the beginning of the section on stationary radar, we established three conditions of relative motion. The three basic situations we mentioned in respect to relative motion were:

1. Radar stationary–The object is moving towards or away from the radar.
2. Radar moving–The object is stationary.
3. Radar moving–The object is moving towards or away from the radar.

In stationary radar, we specified the radar was stationary (situation number 1). But we also mentioned that when the radar is moving, it identifies all stationary targets as moving objects. If, while riding in a vehicle, we were to point stationary radar at a tree along the side of the road while moving, the radar would see the tree as moving towards the radar and the radar would display a speed (situation number 2). Further, if we pointed the radar towards a moving vehicle while the radar was moving, then the sum of both vehicles' speeds would be displayed as motion relative to the radar. I recall using a handheld radar device called a Speedgun while I was a police explorer. Pointing this radar at approaching traffic,[4] I would get both the speed of the target added to the speed of the vehicle I was in. I recall thinking, if I had a way to subtract my speed, I could know the speed of the approaching vehicle. Apparently, the radar engineers had already considered this thought because in 1971, they introduced moving radar. Radar that is designed to detect relative motion while the radar is moving is referred to as moving radar.

First, understand that radars receive many signals at the same time. Stationary radar pointed at multiple vehicles traveling on the highway will receive signals from each, like aircraft radar does. The big difference between aircraft radar and police traffic radar is that it is designed to give the operator only one speed readout.[5] Radar "sees" each target but displays only the strongest signal received. By adding a filter to radar, the ability to isolate the radar's speed from the speed of other traffic becomes possible. Radar gets its speed from the ground directly in front of the vehicle it is mounted in. This signal is close (position) and vast (reflective capability), giving radar a good return signal. Once this signal is established, it will be the lowest speed of all signals present. Because it's the low speed. the signal is termed as the *Low Doppler signal.* All other moving objects will be the speed of the radar AND the speed of the other target combined. Keep in mind that the radar deals in absolute values so there is no regard as to which direction the targets or the radar are moving; either towards or away from each other. The signals from other targets are farther away (position) and have less returned energy (reflective capability) so they are weaker signals. These other signals are higher speeds than the speed the radar unit is traveling, so the signals from these targets are called the *High Doppler Signal.* Even if the vehicle carrying the radar is traveling at 90 mph (144.8

4. I was on the passenger side of the vehicle. I do not advocate trying to manipulate a radar antenna while operating a motor vehicle.

5. There are radars that give more than one speed at a time which will be discussed later.

km/h) and the approaching target is 20 mph (32.2 km/h), the signal for the radar's speed is still the Low Doppler, because 90 mph (144.8 km/h) is less than the combined speed of the radar and the approaching target which is 90 + 20 or 110 mph (144.8 + 32.2 or 177 km/h).

Let us look at a detailed example of how this works:

A law enforcement vehicle is traveling on a highway at 62 mph (99 km/h) with a mounted radar unit in the moving mode of operation. The radar has established a signal for the speed of 62 mph (99 km/h) based on travel over the ground. This signal is processed through the radar Low Doppler filter and displayed this speed in the patrol readout window. It is commonly referred to as the patrol speed readout on the radar instrument. Approaching the radar from the opposite direction is another vehicle. The radar detects the combined speed of the target and the radar as 134 miles per hour (214 km/h). This is identified as a High Doppler signal and filtered. Although both speeds appear in the radar processor the high Doppler is not displayed anywhere on the radar instrument. The processor in the radar subtracts the Low Doppler from the High Doppler. The resulting speed of the target appears on the radar in the target readout window. The mathematical process is identified as the moving opposite direction formula would look like this:

Closing Speed (CS) – Patrol Speed (PS) = Target Speed (TS)

In our example, the calculation would be:

U.S.	Metric
134 – 62 = 72 mph	214 – 99 = 115 km/h

In this way, radar can display both the patrol speed and the target speed at the same time. Early moving radars may have only had one window on the display to allow the operator to see speeds. In a one window situation, the operator had a button which could be used to switch the display from target speed to patrol speed and back again.

When this radar was first manufactured, it consisted of a single antenna which was designed to mount out the driver's side rear window. With the antenna facing rearward, the radar could be applied to target vehicles traveling towards or away from the radar while the radar was stationary. With the radar antenna facing forward, the radar could still operate stationary or it could be operated as moving radar, determining speeds of target vehicles moving towards the radar. Later, the antenna moved inside the law enforcement vehicle and some manufacturers added a second antenna to the radar unit, mounted in the rear deck of the car or on the rear crash bar of a motorcycle. With a second antenna, the radar operator could switch from front to rear antenna operating in either stationary or moving operation. The rear antenna was seldom used in moving operation. The primary reason was in rear facing moving operation, the target vehicle would be traveling away from the law enforcement vehicle. This made apprehension difficult. Of course, the computation was the same. Like in stationary operation, the determined speed was an absolute value so the radar did not know which direction the target was going, only that there was a speed difference.

Just like stationary radar, the relative motion of the target vehicle is subject to cosine, based on position of the radar unit to the path of the target vehicle. And, like stationary radar, this form of cosine is advantageous to the target vehicle. However, there is another cosine to consider. An angle may be introduced to the signal the radar is receiving from the terrain giving the patrol speed to the radar.

As long as the radar antenna is facing in line with the path of the law enforcement vehicle, the radar will display the actual patrol speed. But if the antenna is set to any angle from that path, the same effect in stationary will occur and the radar will read a less than actual speed for the patrol side of the equation. In our example, when the law enforcement vehicle is traveling 62 mph (99 km/h), an angle of just 8 degrees will reduce the patrol speed by 1 mph (2 km/h). This is not to imply that small intentional angles are acceptable. Rather every effort should be made to keep the antenna as in line with the path of travel as possible. The reason for this is apparent when you look at the fact that the target vehicle speed is dependent on a proper calculation by the radar. Let us see what happens when we introduce a 15-degree angle to our 72 mph (115 km/h) vehicle:

The correct computation was CS – PS = TS so that the patrol speed was established as 62 mph (99 km/h)

U.S.
134 – 62 = 72 mph

Metric
214 – 99 = 115 km/h

But, with an angle of 15 degrees, we adjust the patrol speed by 3 mph (or in metric we adjust the patrol speed by 5 km/h). This results in the following calculation:

U.S.
134 – 59 = 75 mph

Metric
214 – 94 = 120 km/h

As you can see, the patrol speed was reduced and the resulting target speed was raised by the same amount. The amount of this change depends both on the number of degrees out of alignment and the actual patrol speed traveled.

Table 6 shows how much of an error can be experienced. This type of cosine effect is called a cosine error since it results in an incorrect reading that is not in favor of the target vehicle. Note the amount of speed loss with an antenna that is misaligned by 20 degrees. The table shows a speed loss of 2 mph at a speed of 30 mph (4 km/h at a speed of 50 km/h), but at the higher speed of 75 mph (120 km/h), the speed loss is 5 mph (8 km/h) for the same 20 degree misalignment.

Some have asked if the moving cosine effect, which occurs due to the angle between the path of the target and the radar, and the cosine error, which occurs due to the angle between the radar antenna and

Table 6.
COSINE AT VARIOUS SPEEDS

	30 mph	45 mph	60 mph	75 mph		50 km/h	75 km/h	95 km/h	120 km/h
5 Degrees	29	44	59	74		49	74	94	119
10 Degrees	29	44	59	73		49	73	93	118
15 Degrees	28	43	57	72		48	72	91	115
20 Degrees	28	42	56	70		47	70	89	112
25 Degrees	27	40	54	67		45	67	86	108
30 Degrees	25	38	51	64		43	64	82	103

the path of the law enforcement vehicle, cancel each other. That would be the case if the angles and the speeds were the same, but they are not. Generally, the cosine effect angle is small while the cosine error angle is large. In all testing that has been done, the cosine error has a larger impact on the target speed, and in a disadvantage to the target vehicle.

Same Direction Radars

Consider the radars that have been discussed up to now and how they are applied.

Stationary Radar–Radar that can determine the speed of a vehicle traveling towards or away from the radar unit while the radar unit is stationary, and

Moving Radar–Radar that can determine the speed of a vehicle traveling towards or away from the radar unit while the radar unit is moving.

For about 30 years, the stationary and moving radars were the only radars used by law enforcement. Then radar engineers came up with a type of radar called same direction radar. As the name implies, this radar determined speeds of target vehicle which were traveling in the same direction as the radar. What the engineers discovered was that by changing the computation method, they were able to allow the radar to make speed determinations in this manner. In order for same direction radar to function, several conditions must be established. In order for the radar to make a speed determination in same direction, there are certain new conditions that the radar must be set to. The relative motion of the radar and the relative motion of the targets allowed are the main consideration.

In opposite direction radars, the slowest motion the radar will see is dependent on what the radar's sensitivity is set to. Some manufacturers set this speed to 5 mph (8 km/h), while others set this speed as high as 15 mph (24 km/h). This also affects the closing and separating speeds. In same direction radar, the radar is made both more sensitive and less sensitive to relative motion. For the radar's movement over the ground, the patrol speed required is generally 25 mph (40 km/h), but the relative motion with regards to target vehicles is set to 3 mph (5 km/h). By setting these limits, the radar can ignore opposite direction traffic. Also, because the radar is computing how much faster or slower a target vehicle is going than the radar vehicle, the radar can detect minor relative speeds. Let's see how this works.

In opposite direction moving radar, the radar vehicle must be moving 10 mph (16 km/h) in our example. Also, the target vehicle must be moving the same speed for a combined minimum speed of 20 mph (32 km/h). If the radar vehicle is traveling at 40 mph (64 km/h), the target vehicle still has a minimum speed of 10 mph (16 km/h) for a minimum closing speed of 50 mph (80 km/h). I think you can see that the speed of the radar vehicle helps set the minimum relative closing speed. As the radar vehicle increases speed, the minimum allowable combined speed also increases.

In same direction radar, the minimum speed is 25 mph (40 km/h), but the condition is reversed causing the radar to block any speed greater than the minimum of the radar vehicle. An approaching vehicle from the opposite direction would be always greater than 25 mph (40 km/h) since the speed of the radar vehicle and the speed of the target vehicle are added to a combined speed. The same direction radar is looking for small relative motion, under 25 mph (40 km/h), so only same direction speeds are shown.

There are two conditions, or modes of operation, which are considered for same direction operation. One condition would be where the target vehicle is traveling faster than the law enforcement vehicle. This would be referred to as a target faster situation. The other would be where the target vehicle is traveling slower than the law enforcement vehicle. This would be referred to as the target slower situation. We will examine the mathematical process used to determine same direction speeds in the target faster situation first as it is the most common application. The formula used for same direction-target faster is:

Patrol Speed + Separation Speed = Target Speed (PS + SS = TS)

Let's examine this formula for a moment. In opposite direction radar, we started with a closing speed (CS) and then subtracted a patrol speed (PS). In same direction, we started with a patrol speed (PS) and then ADDED a separation speed (SS). At this point, it is important to not get confused by the letters CS and SS. These are simply symbols or a place holder where we will put numeric values. Radar that has determined movement regards both closing and separating speeds as the same. It is, to the radar, relative motion or a speed difference. In the opposite direction radar, we are asking "how much of the closing rate belongs to the target vehicle?" In same direction radar, we are asking "how much faster (or slower) is the target vehicle than the law enforcement vehicle?"

We will now examine the mathematical process used to determine same direction speeds in the target slower situation. The formula used for same direction-target slower is:

Patrol Speed minus Closing Speed = Target Speed (PS – CS = TS)

In this equation, we have only changed the sign from plus to minus (and the condition from separation to closing).

Let's see how this would work for both faster and slower target vehicles. We will use a situation where a law enforcement vehicle is traveling at 45 mph (72 km/h) in both circumstances. First, we will compute how the radar determines the speed of a vehicle moving 10 mph (16 km/h) faster. Observe:

U.S.	Metric
PS + SS = TS	PS + SS = TS
45 + 10 = 55	72 + 16 = 88

Now, let's see how the radar determines a speed that is 10 mph (16 km/h) slower:

U.S.	Metric
PS – CS = TS	PS – CS = TS
45 – 10 = 35	72 – 16 = 56

The particular operation desired will often be at the discretion of the operator. In a situation where the operator sees a vehicle traveling faster than the radar vehicle, then the operator can use the radar in target faster mode (this is often the default mode) to have the radar determine the speed of the vehicle. By using target faster, that particular formula is used to calculate the speed. If the operator sees that the target vehicle is being overtaken by the radar vehicle, then the radar mode is switched to a target slower process, causing the radar to use that formula to calculate speed. Traditionally, the same direction radar requires that the operator select which mode of operation to be used.

HIGH-TECH RADARS

Directional-Sensing Stationary Radars

If you were ever involved with citizen band radios (CBs), you would have heard the term "single side band." In a way, that is what direction-sensing radar is. Direction-sensing radar can determine the direction of travel a target is moving as well as having the radar set to only that direction of travel. Now, it is not my intention to confuse the discussions. Up to now, radar returned the change in frequency or Doppler shift as an absolute value, and they still do. What this radar does is determine which side of the transmitted frequency the radar reflected signal returns at. In CB talk, you are said to be operating on the upper or lower side band of a channel. If you are using channel 22 on a CB radio, then there is a channel 22 upper and a channel 22 lower. In radar, it is the same concept.

If the radar transmits at 24,150,000,000 (24.150 GHz) and the reflected signal is changed to 24,150,007,200, then we have determined a Doppler shift of 7,200. In directional-sensing radar, the radar also recognizes that the signal was returned as an upper frequency to the transmitted source. This indicates that the target vehicle is moving toward the radar. If the returned signal was 24,149,992,800, then the Doppler shift was still 7,200, but the returned frequency was on the lower side and therefore the target vehicle was moving away from the radar. These radars are referred to as directional-sensing radars.

In stationary radars, direction sensing can be set to report direction of travel, such as approach or recede (towards and away in some radar) or it may be restricted to that particular direction. If you only wanted to monitor one direction of travel, then the radar can be set to only approach or only recede. Although this does not directly add a tracking history element, it certainly adds to the ability to identify your target.

Moving radars in opposite direction mode are not affected by directional sensing since the target is only moving towards the radar (or away if a rear antenna is used). Assistance in determining or restricting direction is not needed. But in same direction radar, there is a very useful application. Like opposite direction, the use of radar in same direction generally only deals with one direction of travel by the target vehicle. But there is a need in determining direction of travel of the radar in same direction mode.

Recall that the radar in same direction has two modes; faster and slower. With moving same direction–direction-sensing radar, it is possible to determine and even restrict the radar to a faster or slower mode. For example, if the operator wishes to only know the speeds of targets traveling faster than the radar vehicle, the radar can be restricted to only that situation, and slower vehicle speeds would be ignored. But to go one step further, the operator could simply allow the radar to determine direction automatically (Automatic Direction-Sensing Radars). With this situation, the radar will determine the condition the radar is in, receding from a target (faster) or approaching a target (slower), and based on that determination, the radar would then perform the appropriate formula to use and display the correct target speed for all targets detected. In the latter, if the radar detects that the target is going faster than the radar, the faster formula is selected (TS=PS+SS) and the speed is calculated. If the radar detects that the target is going slower than the radar, the slower formula is selected (TS=PS-CS) and the speed is calculated. There is no real change in the audio Doppler and I do not recommend a change in the Target Speed Discrimination Test. It may seem a useless step with the automatic features, but by conducting this test anyway, you can testify that the radar did perform as expected.

Vehicle Patrol Speed Sensors

A Vehicle Patrol Speed Sensor may go by several trade names such as VIP or VSS and is designed to help eliminate cosine problems in radars. An early version was such a device that took the patrol speed from the speedometer, directly eliminating cosine error. But this had problems with improper speed data due to tire sizes and the fact that auto manufacturers would change speedometer designs. Current versions do not replace the radar patrol speed but rather compare the radar patrol speed to the onboard vehicle computer that records vehicle speed information. If these two speeds are not in sync, the radar target speed is blanked.

Another feature to this device is that it can detect when the radar vehicle is in motion. This allows for the device to tell the radar to switch from stationary to moving mode. For example, if the radar vehicle is in moving mode and the operator comes to a stop, the radar will automatically switch to stationary mode. When the radar vehicle starts to move again, the radar switches back to moving mode automatically.

Target Fastest Radars

In the next chapter, you will find the advantages that laser speed detection has over radar. Radar manufacturers attempted a solution to the problems laser solved, with target fastest radars.

It has already been stated that radars receive many signals but only display the strongest. With a target fastest radar the operator may select a temporary condition where the radar can display another signal. Depending on the conditions programmed by the manufacturer, the radar will display a second, faster (or fastest) moving vehicle. The target fastest feature is a stationary function in most cases, although there have been some attempts to incorporate this feature on moving radars. With this radar feature, radar can obtain a faster but slightly weaker signal. For example, if the radar is displaying a large truck at the speed limit, but there is a fast moving vehicle behind the truck, as observed by the operator, the target fastest feature can be applied and the speed of the fast moving car appears in a secondary target window. In this way, the speed of the second vehicle is not lost due to a large vehicle blocking out the signal.

But take care; remember that a FULL tracking history is required. Additionally, this requirement becomes more difficult with the more vehicles on the highway and the number of signals to be processed. It is not too difficult to determine which signal is being received when there are two or three vehicles in range of the radar. But as those numbers increase, so does the difficulty in target identification and tracking history.

Combining Features

Radars in the early years were pretty straightforward. There were few features and deciding what you wanted in your radar was easy. Now there are a multitude of features to choose from. Manufacturers go to great lengths to determine what their customers want. During a class I once conducted, an officer brought his cruiser with two radars on the dash, a handheld radar unit on the passenger seat, and a Vascar unit mounted above the siren and radios. Certainly the desire to have as many features as possible was there. Radars have been manufactured to include time-distance functions, and to combine features such as same direction and direction sensing. One manufacturer combined two radars into one allowing for the radar to function in opposite direction and same direction at the same time. Another manufacturer combined a radar and an in-car video system into one unit. These kinds of innovations clean up the dash of the car and eliminate the number of objects in the cab of the vehicle as well as improve visibility.

RADAR TRACKING HISTORY

We have discussed how radar makes speed determinations. We have also discussed how radar can give incorrect speed readings. How can we be sure that the reading on the radar associates with a particular vehicle and that the reading itself is correct? This is done through a process called tracking history. Of the entire knowledge base in this chapter, tracking history is undoubtedly the most important to understand. Tracking history is a multi-step process used by the operator of the radar to determine an answer to those questions we have asked. Generally, these steps are done in order, and each must be done before a citation can be issued. There are slight variations to tracking history depending on the type of radar being used. We will look at each in the following description.

Stationary Radar Tracking History

Stationary radar tracking history is the foundation to all of the other radar types we will be discussing. The three steps to stationary radar tracking history will be done regardless of the radar type.

Visual Estimation: of the target speed is the way of stating that the operator has made a visual determination of the speed that the suspected target vehicle is moving. In any situation where a traffic violation is suspected, there must be an accuser. Are there situations where the radar is the sole accuser? There are companies that manufacture devices that can determine speeds without human intervention. For this discussion, we are talking about radar units that are not capable of assigning a speed to a single target vehicle without human interpretation. In those cases, there must be a person who will accuse the person driving a vehicle of violating the traffic laws. In the United States, a person has the right to face his or her accuser. To face an accuser is to be allowed to question them. It is difficult to question a device. This has been a common challenge for many years. A polygraph machine and an intoxilizer both have the ability to bring evidence against a person through scientific process but come under suspicion without a person who can explain and answer questions about how this evidence is obtained. Radar is no different. In this step, the operator of the radar must see the vehicle suspected of traveling at an unlawful speed and must assess what that speed is. The operator does not need to be exact, but he/she needs to be very close to the speed the vehicle is traveling. It has been generally accepted that the speed estimate needs to be within plus or minus 5 miles per hour U.S. or plus or minus eight kilometers per hour of the speed the radar determines the vehicle to be traveling. This is a window of 10 mph or 16 km/h. The operator has identified a target vehicle and given that vehicle an assessed speed.

Audio Doppler Confirmation: Radars have an audible sound they make which gives a pitch in correlation with the amount of energy being received. This tone is referred to as the audio Doppler of the radar. This tone has been used for a variety of things in the past. Older radars used the tone to correlate to the speed of the target vehicle and approximate range. In these radars, the closer the target vehicle is to the radar, the louder the tone. For the most part, these other measurements have been eliminated in the audio Doppler. Today, most traffic radars broadcast this tone based on the amount of speed the radar is receiving from the target vehicle. In stationary radar, the tone relates to the speed of the target vehicle. In moving radar, the tone relates to the "combined" or "separation" speeds; the speed that results from the radar's movement and the target vehicle's movement together. As the radar receives the returned signal from the target vehicle, the radar will generate a tone with a pitch that correlates to the speed of the target. This tone differs with radar models depending on which tone generator the manufacturer used. In stationary radar, the tone will be a strong, unbroken resonation with a pitch that is higher in concert with the higher the speed received. For example, if the target vehicle is traveling at 60 mph (96 km/h), the pitch will be heard as a shrill pitch. The pitch will remain steady with a strong returned signal, indicating a good source of the signal. With a speed of 30 mph (48km/h), the pitch would be notably lower, and for a speed of 90 mph (144 km/h), the pitch would be very notably higher.

Radar Digital Readout: once the first two elements of tracking history are established the radar will display a speed on the digital readout of the unit. The expected result is that the visual estimation, the audio Doppler tone, and the readout agree. If they do not agree, then the determination of the radar reading needs to be disregarded. It is important to note that the fact that the tracking history failed does not mean that the vehicle was not driving at an excessive speed, but rather that for some reason the radar did not give the correct or desired information. Operators should endeavor to determine why the radar failed to give the correlated information. This cannot always be done, but most of the time, the operator can make this determination, and in doing so, may be able to make adjustments in location or operation that will enhance better return signals. Possibly because of terrain the radar cannot receive a good returned signal from the desired targets and thus the radar picks up other objects.

For stationary radar, an example of a good tracking history may be:

A radar operator sees that a vehicle which appears to be traveling in excess of the posted speed limit of 35 mph (55 km/h). Watching the vehicle, the operator estimates that the vehicle is traveling between 50 and 55 mph (80 and 88 km/h). The radar generates a tone that is strong and clear and correlates to the estimated speed. The radar then displays a speed of 52 mph (83 km/h).

Recall that the order is important. On one occasion, I was stopped in the median of a highway and while running radar, I was working on paperwork. Suddenly, I heard a very high pitch coming from the radar. Because of the pitch, my attention turned to the radar, which was displaying a speed of 95 mph (152 km/h). I immediately looked down the highway to see a vehicle traveling at a high rate of speed. In this case, I did not do the tracking history in the right order. But let's not get too legalistic with the process and forget the intent. I clearly could make a determination of the fact that the target vehicle was traveling in excess of the posted limit. The process of correlation may happen very quickly. The final determination that the radar reported the correct vehicle speed to the accused driver is a question of the operator's integrity as well as the information provided. The radar may have called my attention to the fact that there was a high speed vehicle in the area, but I still had to visually determine the speed of the vehicle I suspected, and reconfirm the audio Doppler and the radar digital readout.

Moving Radar–Opposite Direction Tracking History

With moving radar the same tracking history as stationary radar is conducted, but there is a variation in audio Doppler and a new element added called Patrol Speed Verification.

Recall that the audio Doppler is either the direct speed from the target vehicle, or it is a result of the combined speeds of the radar and the target vehicle. In moving radar, the audio Doppler is generated from the combined or separation speeds as the situation dictates. If the target vehicle is traveling at 60 mph (96 km/h) and the radar is traveling at 50 mph (80 km/h), the combined speed is 110 mph (176 mph). This will result in a pitch that is higher than for 60 mph (96km/h) in stationary. The behavior of the pitch is the same; the faster the target, the higher the pitch. But it needs to be understood that since the speed the radar is traveling also plays a role in this part of the process, changes in the speed of the vehicle the radar is mounted in also affect the pitch determination. If while operating a moving radar and receiving a returned signal from a moving vehicle approaching from the opposite direction, the vehicle the radar is mounted in increases speed, the pitch will also increase. Care must be taken on moving radar operation to be aware of all conditions of movement in determining tracking history.

One manufacturer, Stalker Radar, eliminates this problem by using a slightly different process. Rather than take the audio Doppler from the combined speed (or separation speed), the audio Doppler determined *after* the radar has computed the target speed. By using an audio Doppler tone after the computation of the patrol and target speed, the resulting pitch is directly related to the target vehicle similar to the pitch from stationary radar operation.

When operating moving radar, we have said that there is a possibility that if the radar antenna is misaligned, the patrol speed may be calculated as lower than actual, thereby raising the target speed. To make sure that this is not occurring, the operator needs to observe the patrol speed readout of the radar and verify that it is accurate by comparing it to another accurate patrol speed source. This is called the patrol speed verification test. This source may be the vehicle speedometer or a Global Positioning System (GPS), or some other speed determination device. To be valid, the device must be certified accurate. This can be done by testing methods and there may be a need for the person who certified it as accurate to be available to appear in court and establish its accuracy along with the test methods used. When verifying the patrol speed, the radar reading in the patrol speed readout location should not differ more than 3 mph (5 km/h) from the other verification device. This step should follow the basic tracking history mentioned for stationary radar and these four steps make up the moving radar, opposite direction, tracking history.

Moving Radar–Same Direction Tracking History

In same direction radars, there is again a change in the expected pitch of the audio Doppler (except for Stalker Radars). Also, since radar units are designed to calculate movement without regard for direction of travel (using the absolute value of frequency change),[6] same direction radar may require operator input to select the proper mode of operation, target faster or target slower. Operation in the wrong mode can give improper readings.

The operation of radar in same direction is a moving type operation. Again, the radar is going to use the combined or separation speed to determine the pitch. Since the speeds used in this type of radar vary more dynamically than in opposite direction radars, the pitch will seem very different. A vehicle traveling at 60 mph (96 km/h) may only be 10 mph (16 km/h) faster than the radar resulting in a very low pitch.

If the same direction radar is being operated in target faster mode, there is an assumption that the target is moving faster than the radar vehicle. But generally, there are a number of vehicles on the road in front of the radar. Any vehicle traveling 3 mph (5 km/h) is not visible to the radar since, to the radar, there is no relative motion. But any

---

6. These radars will become known to you as "Non-Direction Sensing Radars." In the next section, we will discuss those radars that do determine direction of travel.

other moving target is a target to the radar. If, for instance, there are both a faster moving vehicle and a slower moving vehicle in range, the radar could select the returned signal from the slower moving vehicle, even though it is in target faster mode. This results in an incorrect reading for the radar display. Take a radar vehicle speed of 45 mph (72 km/h) and a slower vehicle speed of 10 mph (16 km/h) less than the radar vehicle. If the radar is being operated in target faster mode, the computation (see below) results in a target speed of 55 mph (88 km/h) when the target is, in actuality, moving 35 mph (56 km/h).

U.S.	Metric
TS = PS + CS	TS = PS + CS (Formula Applied)
55 = 45 + 10	88 = 72 + 16

Note the input relative motion is correct but the operative (+) is incorrect.

To remedy this problem the operator will need to be sure the radar is in the correct mode of operation. This is verified through tracking history. The four steps discussed to this point are completed: visual estimation of the target vehicle speed, audio Doppler correlation, digital speed readout, and patrol speed verification. The next step is to conduct a target speed discrimination test. This is done by accelerating the radar vehicle to match the target vehicle speed. While in acceleration, two things will occur. First the digital readout will not change if the radar is in the proper mode (see Table 5.4 A & B), or if the radar is in the wrong mode, the speed display will increase. Second, the radar audio Doppler will decrease in pitch with an increase in radar vehicle speed.[7] If the pitch increases with an increase in radar vehicle speed, the radar is in the wrong mode.

One final comment on the types of radars we have discussed. There are certain consistent characteristics such as all radar operates on the Doppler principle. Also, all police traffic radar sends their signal as a continuous wave. Radars that can tell distance, such as aircraft radar, are generally called pulse radars. At the time radar was developed for law enforcement use, the pulse radars that could judge distance were very bulky and very costly (about 20 times the cost of po-

7. Excluding stalker radars.

lice radar). As time passes, costs come down, and the previous technology becomes available to others. One manufacturer today has radar that will give a range to the target. Depending on patent rights, this may become popular. The truth is that this technology is in a state of constant change and the operators should never take the word of anyone who says it works; always seek to find out how it works.

Chapter 6

RADAR OPERATION AND TESTING

RADAR CHARACTERISTICS AND ANOMALIES

Let's start by considering the features of early radars, and compare them to more modern units. Many of the named anomalies were identified in these early radars and manufacturers have worked to eliminate them. My examples and comments involve radars manufactured in the United States. In many cases, radars manufactured in Europe and Asia followed the same modernization. A good example of the modernization of radar is with the number of readout windows provided. Kustom Signals produced the first moving radar called the MR-7. This radar was similar to their TR-6 stationary radar with a computer unit (often called a counting unit in the 1970s) and a separate antenna. The radar antenna on the MR-7 and TR-6 were quite large and were generally mounted outside the driver's side rear window (see Figure 11). Both radars had a single readout display. But we learned in the last chapter that moving radar requires a verification of patrol speed. The MR-7 had a button on the computer unit that bore the letters "VFY", indicating "verify." When a target speed was displayed, the operator would press the "VFY" button and the target speed would change to the patrol speed. It was soon discovered that the process of switching from target to patrol speed opened a question of whether or not conditions changed during this process. The operator could not testify that the patrol speed observed in the radar display was the patrol speed when the target speed was displayed. Other manufacturers with one window radars, such as Decatur with their MV-715, saw the problem and they changed their radars to two-window radars.

Figure 11. A 1972 radar antenna.

Early radars allowed the operator to manually lock in a speed, or to establish a threshold speed so that the radar could automatically sound an alarm sound and lock on any speed above the set threshold. This caused two problems. The first was what was commonly called the "comic book syndrome." The allegation was that the operator would be reading instead of monitoring traffic. The radar would sound an alarm indicating that a vehicle was traveling in excess of the established threshold. The operator would look up and see a vehicle thought to be speeding. It was often questioned in court whether the operator had an opportunity to establish a tracking history or if it was a reaction to view the first vehicle in sight as the speeder. The second problem came from the fact that once locked, the tracking history was terminated, and additional information on target identification was eliminated. Progressively over the years, automatic alarms and locks were eliminated. Another feature of early radars related to locking speeds was the ability to "hold" a speed reading. Many types of radar were manufactured with the ability to keep a speed reading on the display for a couple of seconds. If the operator saw a displayed speed, this feature would give a few seconds to lock in the speed. At times, this created confusion, especially when the radar was moved from one target to another.

Common Radar Anomalies

When using radar, we must understand that the device is sensitive to movement. Movement can be introduced to the device that can cause the radar to give unintentional speed readings. In addition, radar can receive other radio waves, which can give readings that in the past have been referred to as "ghost readings." All these have been called errors but in fact are not errors at all. The readings occur based on physical laws and because certain conditions have been established which give the radar information that is believes to be reliable. The actual operation may be improper, but the resulting readings are not erroneous. For that reason, I have elected to use the word anomalies to describe these situations.

It is important to recognize that in every anomaly stated, a speed reading displayed would not meet the conditions of tracking history. That is to say that with tracking history, the operator can readily identify that the radar is experiencing an anomaly. Conversely, without tracking history, there is no way to be sure the speed reading is actually not an anomaly. In each of the anomalies discussed, the first element of tracking history, visual estimation of speed would not be satisfied. In the rare instances where it is thought that the speed displayed was that of a vehicle you estimated, then the audio Doppler will lack a consistent correlation. Very often the audio signal will be broken and weak. And, anomalies are often brief in their duration, and they are somewhat weaker signals when compared to actual moving targets allowing them to vanish when a strong target reading appears.

Do not use these rules as written in stone. I have seen certain anomalies that are strong and persistent. Actually these strong anomalies reveal themselves because they are inconsistent with the target behavior. The display is not in sync with vehicle movements which should raise a suspicion that the radar is experiencing an anomaly. It is best to always ask yourself why the radar is displaying any reading. In this way, you will become comfortable in identifying valid targets from anomalies.

Radar that is operated in stationary mode is designed to operate while the radar is in a fixed position or otherwise not moving. If the radar is moved in this mode, unwanted readings may be produced. We have previously discussed that stationary radar pointed down a road will read the movement of the radar and at times can give the combined speed of the radar and an approaching vehicle in the display. One of the more common situations may occur when the radar is being operated handheld and is allowed to move while it is transmitting. While operating the radar, the operator may change the direction it is pointed and the rapid movement from one point to another can produce a speed. This anomaly is referred to as *scanning*. This situation was more of a problem in early radars due to the fact that early radars "held" the speed reading for a few seconds to be sure the operator did not lose the opportunity to lock the target speed just acquired. This could allow the operator to confuse the readout with respect to which car was traveling the indicated speed. Also, if at the conclusion of rapidly moving the radar a moving vehicle was encountered, the held speed could be added to the target speed resulting in a speed higher than actual.

Radars are manufactured to either be held in the hand using a gun-type frame, or to be mounted in a vehicle with an antenna separate from the computer unit. Some radar can be used as both; the Decatur Ra-Gun was handheld radar in which the antenna could be detached, cabled together, and used as a mounted unit. Any radar where the antenna is not a part of the computer unit has the potential to interact with itself. Operators discovered that when the antenna was pointed at the counting unit while transmitting, a variety of sounds similar to an arcade game would be produced as well as a variety of speed readings. This unintentional action is referred to as *panning*. The speed readings produced are generally very sporadic and unattributed to any moving object. The fact remains that if the antenna is mounted in proximity of the front of the radar, the anomaly can be produced.

When moving radar was produced in the 1970s the ability of the radar to compute a frequency change was done using analog processing. It is not necessary to get into the electrical engineering aspects of radar, but we should understand when talking about a processor (the ability to process information) that analog describes slower and larger electronics. As radar receives information about both the patrol speed (speed over the ground) and a target speed (the combined speed of the radar and another vehicle), a filter sends this information to the appropriate place in the computer unit. This filter is like a turnstile or revolving door. If that door is not turning fast enough for information changes, some of the data is lost. This is a limitation of the radar. As the radar vehicle is accelerated or decelerated rapidly, there is the potential for the radar to lag behind in information. This causes the radar to report either a lower than actual or higher than actual patrol speed, an anomaly known as *batching.*

The result from batching is a miscalculation of the actual target speed. Like cosine, batching may report a lower than actual patrol speed. Inserting this fact into the formula for moving radar opposite direction will result in a higher than actual target speed. There is a potential for the inverse to occur. If the radar vehicle is being decelerated rapidly, the patrol speed on the radar may fail to keep up with the change, resulting in the radar reporting a higher than actual patrol speed and producing a lower target speed reading.

With the advent of digital signal processing (DSP), modern radars can handle this information much faster. Most radars now process signal information many times a second, preventing batching from occurring. Even so, radar operators need to be aware that the potential still exists. High performance vehicles and motorcycles have acceleration potentials that may still allow for batching. In addition, lost signals due to terrain changes can cause the radar to fall behind briefly on processing a signal. Correlating display information with good tracking history will prevent such an anomaly from becoming a problem.

The *shadowing* anomaly occurs when certain conditions are established. Radar uses the ground to establish a patrol speed. However, there may be other objects around that give a better returned signal to the radar than the ground. For instance, the back of a truck is a better object in reflective capability than the ground. This is fine as long as the truck is not moving. But, if the truck is moving at a speed lower than the radar vehicle, a miscalculation of the patrol speed can occur. Let's take a look at an example:

A radar vehicle is traveling at 45 mph (72 km/h) when it approaches the back of a large truck traveling 20 mph (32 km/h). The radar signal switches from the ground to the truck, but since the truck is moving, the radar sees the patrol speed as the speed at which the radar vehicle is overtaking the truck (relative motion); a speed of 25 mph (40 km/h). Observe how this change affects a target speed reading of 55 mph (88 km/h) when we use the moving opposite direction formula:

$$TS = CS - PS$$

The actual speed of the target would be displayed with the radar receiving a signal from the ground for the patrol speed.

U.S.	Metric
55 = 100 – 45	88 = 160 – 72

However, when the truck becomes the source of the patrol speed display, there is a noted change in the target speed.

U.S.	Metric
75 = 100 – 25	120 = 160 – 40

Observe that there is no change in the combined speed of both vehicles. The anomaly only pertains to the patrol speed readout of the radar.

Shadowing can only occur when certain elements are in place. First, the radar must find a more reflective source than the ground. The ground is a large target and a good signal, but lacks reflective capability. The back of a larger vehicle or a length of fencing can produce a better return signal. Second, that source must be in motion at a speed greater than the lower limits that the radar is set to detect. (i.e., greater than 10 to 15 mph or 16 to 24 km/h). Finally, the radar vehicle must be overtaking the reflective source at a speed greater than the lower limits that the radar is set to detect (i.e., greater than 10 to 15 mph or 16 to 24 km/h).

Generally, this anomaly is only present as long as these conditions occur. But, there are circumstances where the radar can continually display shadow speeds. For example, if the roadway has a layer of water on it, then the radar cannot acquire the ground as a good source for the patrol speed. In this situation, shadowing may occur on a continual basis.

Interference is the unintentional reception of a frequency, either mechanical or electromagnetic, which either prevents the radar from displaying a speed or causes the radar to display a speed not related to an actual target vehicle. Some of the more common sources of interference from outside the radar vehicle are:

Electric motors such as power generators–These devices have moving parts and as such can be detected by radar in close proximity.

Power lines and power transformers–Power lines transfer power in cycles, and radar can see these cycles as movement. Also there are stronger signals coming from power lines (which is why you sometimes get static on your car radio), which can give radar interference.

Microwave transmitters such as weather radar–These transmitters pose the least problem since most are pointed away from the ground. If you are close enough, you may pick up some interference, but it is rare.

Harmonics–This problem comes from a revolving object, most common from radar antennas mounted on police motorcycles in the vicinity of the rotating tire.

Mercury vapor lights–Generally you need to be directly under the lights to experience this problem.

Many of these forms of interference can be identified and thus remedied by adjusting the radar sensitivity control. This control, often referred to as the "range" control, is basically an RF gain control. By adjusting this control, the radar can be made more or less sensitive to received signals. If you are experiencing interference from power lines nearby, you may wish to adjust the sensitivity control and see if you can eliminate the interference. Be careful though as you will also eliminate target reading from some distant targets. This becomes a value judgment on your part to give up some range (thus the name of the control) and eliminate the interference. Additional resolution to these forms of interference is to identify the probable cause and remove the radar from the vicinity of the source. Also there are forms of interference which are generated from inside the radar vehicle. These include:

Police transmitters–Often, theses radios have their antennas mounted on the trunk of the vehicle, and due to their high power outputs, they can feed directly into the radar antenna, especially if they are mounted facing rearward.

AM/FM Radio speakers–Listed here, but only a problem with X-Band radar equipment, and very rare.

Vehicle heating and air conditioning fan motors–A consistent problem when the radar antenna is not mounted over them. Also this can be a problem if the vehicle manufacturer has mounted these fans outside the firewall.

In-vehicle computers–Many computers on initial transmit will send a power surge and this can be read by the radar.

Electrical noise from the vehicle electrical system–Backfeed and poor grounding can cause radars to have reception problems. By wiring the radar past the fuse block and directly to the battery can elevate this problem.

You can use the sensitivity control or you can reposition the radar antennas. The electrical noise can be eliminated by building a power source directly from the vehicle battery instead of running through the vehicle electrical system. Each of these anomalies produces a weak signal that can easily identified through the audio Doppler sound. Also they lack tracking history elements. Many times they are short-lived and can be eliminated through repositioning the radar vehicle or radar antenna.

Jamming and Detection

Jamming is not an anomaly because it is intentional. An anomaly is defined in Webster's Dictionary as something abnormal or peculiar. In this book, we have identified radar anomalies as readings which occur based on physical laws arising out of unusual circumstances. But jamming is an intentional attempt to, at minimum, block radar from operating. Radars are considered by the United States Federal Communications Commission (FCC) as "radio location" devices and there are laws against intentional interference. But the FCC has very little staff with a large responsibility. If the violations are of a magnitude to involve the FCC, then the consequences can be significant. But it is rare for them to be involved in single incidents and as you will see, most jammers do not work.

Jamming devices and methods are often the result of rumors or snippets of information without credibility. During World War II, attempts were made to jam radar signals by dropping strips of foil called "chaff" from aircraft being tracked by ground to air radar[1] and motorist have tried using this to block the radar signal. The problem is that the large amounts of foil needed to continually block a potential signal would make this method cost prohibitive. And, as a California Highway Patrolman once said, "I doubt it would work, and it would probably end up in a littering ticket." Here is a list of other home

1. See Chapter 5: History of Radar.

remedies that have been tried with some explanation of why they would not work:

Hanging reflective objects from the vehicle mirror
Tested by Mythbusters,[2] this attempt failed. At best, it probably would have made the vehicle more reflective.
Hanging chains from the vehicle
Covering rims or entire vehicle with foil
Placing metal objects in the hubcaps
Broadcasting a microwave signal from the car
Rapid horn honking
Using a radar absorbing bra on the front of the car
Using radar absorbing car wax

Many of these attempts were tested by "Mythbusters." Mythbusters tried painting a car nonreflective flat black, and with other substances. All tests showed these methods to have failed. When I heard of the radar absorbing car wax (and actually seeing a wax product once claiming to defeat radar), I had to ask, if this method did work, why does the government spend billions on stealth aircraft? All they have to do is wax them.

I was telling a good friend of mine about the methods motorists use to try and jam the radar, and he told me that he knew of a product that can beat the radar; he said, "your speedometer will work every time!"

Radar detection is a close kin to radar jammers. It is not really an anomaly, and it is not interference or jamming, but radar detection is viewed by the motoring public as a way to "beat the radar." The first detectors were called "fuzzbusters" and they have changed little since their appearance in the market place. They simply receive radio signals and broadcast an alert to the user that there is radar in the area. Radar started out operating at 10.525 GHz frequency, so the detectors would look for a signal on that frequency and alert the user. Then manufacturers started to produce radars at 24.150 GHz, and detector manufacturers had to add this frequency.

Radar detectors became very popular in the mid 1970s; so popular in fact that a song[3] was written about them and states passed laws

2. Mythbusters Episode 18, aired October 13, 2004.
3. "Convoy" by C.W. McCall, 1975.

against them. Some regarded the radar detector as a device bent on allowing motorists to violate speed laws. Today, only two places in the United States have laws against radar detectors in noncommercial vehicles: Virginia, and Washington, D.C. On the federal level, radar detectors are not allowed in commercial vehicles and on military bases. In Canada, the only three provinces that allow their use are Alberta, British Columbia and Saskatchewan. Even some European countries outlaw their use such as Spain and Finland.

Despite the view that they are a tool to break the law, many see the radar detector as a tool to reduce speed. Certainly the recognized reaction to receiving an alert from a radar detector is to slow down. But law enforcement also recognizes that such a forced deterrent is short-lived and sporadic at best. The motorist will slow down for as long as there is an alert on the radar, and then their speed will climb up once again. There is a strong belief among law enforcement officers that the best deterrent is a citation. The radar manufacturers also want the operator of radar to have the advantage in speed enforcement. In attaining this goal, the radar manufacturers have created add-on features to radar in order to "beat the radar detector."

The first method to defeat radar detectors was to add a pause control or "hold" switch to the radar. This allowed for interruption in the transmitted signal of the radar. By placing the radar in hold, the motorist with a radar detector did not have the ability to know that they were in the vicinity of a radar unit. Once the radar hold was turned off, the radar signal transmitted immediately. Even if the driver was quick to slow, the radar generally had the speed displayed.

There were advantages and disadvantages to this hold feature. One of the advantages to the hold feature was that the radar vehicle regained a degree of stealth. Criminals had learned to have a radar detector in their vehicle while committing crimes. In this way, they would be aware of the presence of law enforcement. By normally running with the radar in hold until a speeding vehicle is spotted allowed officers to patrol undetected. A disadvantage to the hold feature was that in moving mode, the radar would take a few seconds to acquire the ground for a patrol speed. Often in those few seconds, the radar would display the combined speed of an approaching vehicle and the radar vehicle in the patrol window. This would clear relatively quickly, but while it was being displayed, no other speed could be detected.

Figure 12. A common radar detector.

During this time, Stalker Radar applied to the FCC for a frequency range for new radar they had designed. This frequency range of 33.4 GHz to 36 GHz was granted and allowed the company to manufacture radars at frequencies that would be hard for the radar detectors to determine the correct frequency. Radar detector companies, however, came out with a radar detector that scanned all of the frequencies. Scanning-type detectors are more expensive so there are fewer of them, even today. A recent advertisement indicated a radar detector that scanned all bands: S, X, K, Ka, and Ku. All these bands are assigned police radar frequency bands, but you probably don't recognize all of them since some are no longer used (S-Band) and some are exclusively used in Europe (Ku-Band).

One other manufacturer used a feature to help officers detect speeders over their radar detectors. The MPH Industries radars had a feature called the POP as an option. This feature allowed for the operator to activate the radar for a very brief time. Radar processors are often much faster than a radar detectors and the operator can obtain a speed reading of a target very quickly. Even if the radar detector obtains a radar signal, it appears as a chirp giving the impression that it was some sort of interference. Radar detectors are subject to the same interferences that radars are subject to so the motorist may totally disregard this alert. The operator cannot take enforcement action on this small bit of information, but based on that information, the operator may engage the radar so that tracking history can be completed. The nice thing about both the hold and the POP features is that once the motorist is in the radar beam, their reaction is to slow (sometimes dramatically), and the correlated speed drop on the radar sets positive identification that the displayed speed is the vehicle that is now slowing.

To Lock or Not to Lock

To lock or not to lock, that is the question. Ok, it's not Shakespeare, but it is a question that I would like to end this section with. In the early days of radar, law enforcement got into a habit of showing the violator their speed on the radar. There was something about seeing that speed that gave the violator a feeling of defeat. Was it needed or required? No! But many officers would not even write a citation without it, and motorists came to expect it. On one occasion, I was riding with an officer when we received a speed reading on our handheld radar at 60 mph (96 km/h) in a posted 40 mph (65 km/h) zone. As we started to pursue this vehicle, an accidental bump of the radar caused the reading to unlock. The officer immediately instructed me to hold the radar out the window. Pointing the radar at the side of the road, he accelerated until the 60 mph (96 km/h) was in the display and the radar was relocked. Was the reading in the radar valid? No. Was the motorist still speeding? Yes. In court this would have been an embarrassment. Is it necessary to lock in speeds? No. I admit that it is nice to have that reading to show the argumentative motorist. You really want them to understand that they were speeding. But does it really prove anything?

RADAR TESTING AND HEALTH CONCERNS

In the 1970s, with the advent of moving radar, sales increased. Law enforcement agencies saw that traffic could be monitored and the officer could remain on patrol. Also in 1973, there was a huge emphasis in the United States on a 55 mph speed limit. This brought about public concern. More and more people were being cited with this new device that they had not heard about before. To the public, it was a magic box. They did not understand it so they went to court. Now the courts were being asked if this was a valid way to determine speed. They, too, wanted it explained.

At this point, training was left to the manufacturer, and they had to rely on their sales staff. Some were very good and some were not. And training was very much abbreviated. What the operators thought the radar could do was at times at odds with what it could do, and even the sales staff was not sure what it could do. I recall in the 1970s helping an officer "aim" the antenna. One sales representative told an officer the VFY button allowed the radar to clock a car in front of you. Needless to say, there were some concerns as to what radar was capable of. In 1976, the International Association of Chiefs of Police (IACP) passed a resolution asking the federal government to establish testing criteria for radar. In 1977, the National Highway Traffic Safety Administration (NHTSA) entered into an interagency agreement with the Law Enforcement Standards Laboratory (LESL) of the National Institute of Standards and Technology (NIST) to conduct certain testing of radar devices. This testing, adjusted for current models of radar, is still in existence today.

Testing is now done in most countries throughout the world in one form or another. Some testing is very involved while others may not be so involved. Testing is not an unusual part of the electronics industry. Most professional companies worldwide will have their electronic equipment tested and calibrated regularly. Even the test instruments used to test electrical equipment are regularly tested and calibrated. For radars, this testing is not mandated by law, but it is required if the radar is to be purchased using federal tax dollars. Other testing exists in places outside the United States.

The IACP has a list of approved radars. This list is referred to as the CPL or Conforming Products List. Radar manufacturers who want their radars on this list must submit a certain number of radars for testing and pay for them to be tested. This testing is equivalent to the types of testing conducted by Underwriters Laboratories. Every specification of the radar is tested, and these specifications must meet the minimum performance standards as published by the IACP. These standards are not arbitrary. They were decided upon through the cooperative effort of the National Institute of Standards (NIST). The tests involve laboratory bench testing and field testing. After approval and placement on the CPL, these manufacturers are subject to periodic retesting of their radars in the future.

The radar that is used by law enforcement comes from the manufacturer with a series of tests that are recommended. The frequency of these tests varies sometimes, but in the United States, case law has indicated that they need to be tested at least twice daily, at a point before and after enforcement has taken place. The testing is based on the radar design. There are basic sets of tests for all radars, a set of additional tests for moving-opposite direction radar, and yet another set of additional tests for moving-same direction radar.

Testing Stationary Radar

All radars have the stationary function built into them. At this point, no one has made moving only radar. Therefore, the following tests are required of ALL radars. Note that some radar units have a power on series of tests that run automatically when the device is powered on. This feature does not preclude that all of the tests here should not be run manually.

Light Test: this test is designed to assure that all of the lights on the radar are working properly. This includes the speed display(s) and all warning lights. This test MUST be conducted first for a failure in any of the illuminated portions of the radar display could affect the results of other tests. In older radars, the display consisted of a vacuum tube with elements that glowed. This was later replaced by light emitting diodes (LED) which are used in many types of radar still today. The other type of display is the liquid crystal display (LCD) which is also very popular. Because of the high reliability of the LED and LCD displays, failures are rare; but if there is a failure, the radar needs to be sent to an electronics repair facility. In any case, a failure of the light test would require the radar be removed from service until the problem is corrected.

Internal Test: there are a variety of internal tests depending of the manufacturer's design. Early radar had an internal tone generator which was set to a certain number of cycles per second. Upon activation of this test, the radar would read the internal tone and display a corresponding speed. For most X-Band radars, that speed was 64 mph (102 km/h) and for K-Band radar, the readout should be 32 mph (51 km/h). With the advent of digital signal processing (DSP) and other features, it was found that a series of tests would be used and the information about those tests would be displayed instead of a speed. Some radar would run through a series of numerical values, or give verification of software versions installed. At times, a simple "Pass" displayed or for some radar the absence of an error indication was all that was needed to pass. For the internal test, it is imperative that the operator is aware of which indications (or lack of indications) would be required. The radar MUST display the exact information required for this test. Failure of the internal test would require the radar be removed from service until the problem is corrected.

External Tests (Optional): Once you have completed the light test and internal test, the major testing is actually done. These tests are specified in the owner's manual and it is imperative that you perform all the tests indicated by the manufacturer. One series of tests often specified is an external test, sometimes referred to as the tuning fork test(s). This is a method of artificially introducing a Doppler shift to the radar and determining if it can display an expected speed. Tuning forks are cut aluminum and when struck will resonate with a pitch based on the length and thickness of the fork. Most of these forks are stamped with the frequency band they are made for, and the speed they should display. Let's see how this would work:

> A tuning fork is provided with the radar and it is stamped to oscillate at 4,716 cycles per second, or 4,716 hertz. It is also stamped for K-Band Radar and indicates an expected speed reading of 65 mph (104 km/h) in that band. By striking the fork on a material that is softer than aluminum (plastic works well–and in today's cars you should have no problem finding some plastic), you place the fork about 3 to 5 inches (8 to 13 cm) away from, and in front of, the antenna to the radar. The fork should be turned sideways so that only one tone bar is exposed to the radar. The radar should respond to the fork with a speed displayed at 65 mph (105 km/h). The allowable readout range is 64 to 66 mph (or 104 to 106 km/h). This is because shifts in the tuning fork frequency occur. In my example, I selected a tuning fork frequency of 4,716 (which may not be unusual). If you divide by the Doppler shift of 72 cycles per second, the resulting speed is 65.50 U.S. (In metric the divisor would

be 44.75 and the result would be 105.38 km/h[4]). Changes in temperature can shift the tuning fork frequency slightly. With a fork set to 65.5 U.S., a cold fork may shift down 6/10ths of a mile per hour causing a speed reading of 64 (due to truncation) and a hot fork may shift up 5/10ths of a mile per hour causing a speed reading of 66 miles per hour. In metric, the same effect can occur with just a potential drop of 4/10ths of a kilometer per hour or an increase of 7/10ths of a kilometer per hour creating a speed change of 1 km/h).

As you can see this process is not the most accurate. Also take note that the tuning fork must be used in conjunction with the correct frequency band. For example, if the tuning fork was set for 4,680 hertz (exactly 65 mph in K-band or 104 km/h metric) and held in front of an X-Band radar, the reading of the radar would be 149 mph (238 km/h). The reason for this testing is explained in the case law section of this book, and such case law is based in the United States. Some countries such as Canada do not subscribe to the tuning fork test, but if the manufacturer specifies it, then they may come under attack for it in their courts if they do not do it. I have termed this test as optional because the case law to establish it is now half a century old, and the technology of radar has leaped over the tallest buildings to today's radar technology.

Testing Moving–Opposite Direction Radar

All the tests described in the stationary radar testing section would apply to moving radar used in the opposite direction mode. There is an additional external test (also optional).

External Tests (Optional): If the external test described for stationary radar is conducted, then a second external test for moving-opposite direction would need to be completed. This test involves using two tuning forks designed to oscillate at two different speeds. The point of this test is to simulate both the high and low Doppler signals received by the radar. These tuning forks would be supplied with the radar and are set by the manufacturer. For our example, we are going to describe this test using a 65 mph (104 km/h) fork as we did for stationary and

4. There is a small difference in the mathematical conversion due to the truncation of these speeds.

an additional fork set to 35 mph (56 km/h). In this case, we would start with the low Doppler fork, striking and holding it as described in stationary radar. The radar would be set to moving mode, and the tuning fork would produce a speed of 35 mph (56 km/h) in the patrol speed readout window of the radar. Now, while still holding the first tuning fork in front of the antenna, and while it is still oscillating, strike and hold the second tuning fork in the same manner. The radar should see the first fork as a low Doppler speed and place that speed in the patrol window display. The second fork will be viewed by the radar as a high Doppler or closing speed. The processor of the radar would then use the information provided by the two tuning forks and subtract the low Doppler from the high Doppler (TS=CS-PS). The result would be a speed of 30 mph (48 km/h) in the target speed readout window. Recall:

U.S.	Metric
30 = 65 – 35	48 = 104 – 56

The resulting speed readings must be within plus or minus 2 mph (2 km/h) for the same reasons described in stationary radar testing. The reason for 2 miles per hour (2 kilometers per hour) being allowed is because you are now using two tuning forks which each could experience a shift in frequency.

Testing Moving–Same Direction Radar

All the tests described in the stationary radar testing section would apply to moving radar used in the same direction mode. There is an additional external test (also optional) which is in addition to the opposite direction test.

External Tests (Optional): If the external test described for stationary radar is conducted, then a second external test for moving-opposite direction AND a third external test for same direction would need to be completed. This test involves using the same two tuning forks designed to oscillate at two different speeds as in the opposite direction testing. The point of this test is to simulate both the high and low Doppler signals received by the radar. For our example, we will use the same two tuning forks that were used in the external test for opposite direction radars: a 65 mph (104 km/h) fork and an additional fork set to 35 mph (56 km/h). In this case, we would start with the high Doppler fork, striking and holding it as described in stationary radar. The radar would be set to moving-same direction mode for target faster operation. The result would be 65 mph (104 km/h) in the patrol speed readout window. Now, while still holding the first tuning fork in front of the antenna, and while it is still oscillating, strike and hold the second tuning fork in the same manner. The second fork will be viewed by the radar as a simulation of a target separation speed. The processor of the radar would then use the information provided by the two tuning forks and this time add the separation speed to the patrol speed (TS=PS + SS). The result would be a speed of 100 mph (160 km/h) in the target speed readout window. Recall:

U.S.	Metric
100 = 65 + 35	160 = 104 + 56

The resulting speed readings must be within plus or minus 2 mph (2 km/h) for the same reasons described in stationary radar testing. The reason for 2 miles per hour (2 kilometers per hour) being allowed is because you are now using two tuning forks which each could experience a shift in frequency.

Beyond the Basics

As I stated, the basic tests are sufficient for all radars. You may have observed that we did not test for moving, same direction in target slower mode or for target fastest mode. This is because the testing can get involved. Button and controls on the radar for these two modes require constant attention (they need to be depressed and/or held down) and in some cases they require the use of three tuning forks (I only have two arms). Because they are impractical, the manufacturer does not identify tests for these features. In directional sensing radars, you have a different problem in that they cannot read a tuning fork while in directional mode. Some manufacturers have established methods to conduct tuning fork tests with the directional sensing process temporarily turned off. But is this a true test of the functionality of the radar?

The IACP tests are very involved and complete. In the directional radars, they have a simulator which was built to create a directional condition electronically. It is attached to an anechoic (anti-echo) chamber for the tests they conduct. There are only a few of these simulators in existence making them very cost prohibitive for local use. All the tests conducted by IACP have proven that radars have a very high degree of reliability. I had the opportunity to speak with Dr. P. David Fisher of Michigan State University who is responsible for developing the rules and testing criteria for radars in the United States. He has tested more radars than any man alive and I asked him in all those tests he had conducted if he had ever observed any radar give a false speed. His response was "less than a handful" (implying a very small number). If you are a basic user, or an instructor, this is the limit of what needs to be known. If you desire more information, seek it. There are people in the industry who can use fresh ideas. You may be the one to design the next generation of radar.

RADAR AND YOUR HEALTH

In a world where radio transmitters are so commonplace, no one really gave a lot of consideration to the potential dangers of police traffic radar transmissions. We take the fact that we are constantly bombarded with radiated power in stride. Probably since we cannot see or feel this radiation, it is ignored. And, in reality, it is mostly unknown. But it is still there and it is something to be considered.

Radar Radiation Risks

Radar is a form of energy and it is known that radiation can alter the state of certain organic materials. A microwave oven, for example, has the ability to heat food. What then does police traffic radar do to the human body? There are a couple of things to remember when trying to compare police traffic radar to a microwave. First, consider that you place the food in a chamber that is designed to trap the microwaves and concentrate them. Second, and most importantly, microwaves use more power than police traffic radar, much more power. A microwave oven uses electronics to turn the normal supply voltage of 115 volts into a useable 3,000 volts which allows the oven to produce output power levels of between 200 and 1,000 watts. The police traffic radar operates at less than 5 millionths of a watt in output power.

Even at these low levels, very little is known as to the long-term effects of this radiation. To determine if this radiated energy is harmful would require placing a test subject into the radiated signal path of police radar for at a minimum of 2,000 hours of exposure just to see the potential damage an operator could experience in one year of continuous operation. This is not to say the testing isn't important, but even the hypothesis that police radar is harmful is inconclusive at this time.

In 1989, State Trooper Gary Poynter published an article concerning cancer and radiated power from police traffic radar. Trooper Poynters' study invoked a variety of reactions from both law enforcement and the legislatures in the United States. His study, although not fully scientific, was compelling. Was there a danger and if so, was law enforcement and the government doing everything it could to protect the radar operators. His study got the attention of Senator Joseph Lieberman and a congressional hearing was held in 1992 and this group published a technical report concerning their findings.[5] Unfortunately, very little was learned concerning the effects of long-term exposure. However, several suggestions on safe use of police traffic radar did emerge from this report.

The goal in maintaining your health and safety is to reduce the risk of radiation exposure. But is there a risk at all? If you walked into a kitchen and wanted to know if the stove burners were hot, there are several steps you would take to make this determination. At the conclusion, you would know if the stove burners were safe to touch or not. Prior to these steps being done, you would not place your hand on a burner. With radar, it is the same. You do not know if there is a correlation between the radiated power from radar and the development of various types of cancer. So before and during radar operation, you would exercise certain safety precautions to make sure your health is protected.

Here is a list of recommendations to assist the radar operator in the safe operation of radar devices. Several recommendations, specified in the technical report, pertain to the type, operation, placement, maintenance, and proper usage of traffic radar devices.

---

5. Occupational Exposure to Police Officers to Microwave Radiation from Traffic Radar Devices, U.S. Dept. of Health and Human Services, 1995.

The following procedures are recommended to reduce or prevent exposure to microwave radiation emitted from traffic radar devices:

Handheld devices should be equipped with a hold switch requiring activation to emit radiation. This switch should be designed to allow transmission only when activated, and should automatically return to the hold status. All radar devices should be kept in a hold status when not monitoring traffic.

Radar devices should not be placed with the radiating antenna pointed toward the body, whether it is held in the hand or placed near the operator. A holster or other similar device should be used as a temporary holder for handheld radar when not in use.

When using two-piece radar units, the antenna should be mounted so that the radar beam is not directed toward the vehicle occupants. If possible, the antenna should be mounted outside the vehicle altogether, although this may not be practical with all units that cannot withstand adverse weather conditions. Other options, e.g., mounting on the dashboard of the vehicle, are acceptable if the antenna is at all times directed away from the operator or other vehicle occupants. The antenna should be mounted as close to the windshield glass as practical.

Radar antennas should be tested periodically, at least annually, or after it had experienced damage, for radiation leakage in a direction other than that intended by the antenna design.

Each operator should receive training in the proper use of traffic radar before operating the device. This training should include a discussion of the health risks of exposure to microwave radiation and information on how to minimize operator exposure. The Institute of Electrical and Electronics Engineers publishes the standard for safety concerning exposure to radiated energy. This standard specifies the maximum limits of radiated output that a human may be exposed to from radar devices. This limit is set in miliwatts per centimeter squared. Over the course of the last 20 years before this book was published the standard was changed three times. Both the radar manufacturers and the IACP are concerned about safety when operating radar devices. It is suggested that if you have an interest or a concern, that you contact either the radar manufacturer or the IACP to determine the tested level of your radar.

As stated, very little is known concerning the effects of cancer and police traffic radar. But, careless operation such as resting handheld radars between your legs, or placing the radar antenna so that an operator is in the beam path during normal operation certainly tends to justify the allegation that it does have a potential relationship. And radar instructors need to be even more aware of these safety precautions and make a concerted effort to see that all of their students understand these principles.

Chapter 7

BASICS FOR POLICE TRAFFIC LASERS

Standing in a classroom after a session on police traffic radar operation, a student asked me what I thought would be the next technology. I gave this question some thought, and then told him that lasers will probably be the next development in speed detection. It seemed logical considering the similarities of radio waves to light waves and the advantages of lasers' pinpoint accuracy. If only I had done some investing in the stock market on my suggestion.

Lasers used for speed detection come with a variety of names: Laser Radar, Lidar, and Laser-Based Speed Measurement. Some of these names have a type of marketing approach to them, and some are claimed to be proprietary by some manufacturers. For our purposes, this book will simply refer to them all as lasers. Before we can look at lasers used to determine speed though, we need to have a good understanding of what lasers are.

THE HISTORY OF LASERS

Lasers were a reaction to the many years of science fiction stories concerning guns that shot beams of light. Oddly enough, today lasers have been applied to many products but weapons applications has yet to been finalized (at least in the way the science fiction writers visualized lasers).

Two men are credited with the development and discovery of the laser. Gordon Gould, a doctoral student who had studied under Charles Towne, developed the idea of the laser. In fact, he was the first person to actually apply the word. In 1958, Gould had completed the work on the laser, but he filed for his patent after Theodore Maiman who had built a working laser and demonstrated it in 1960 at a news conference. At that conference, Maiman stated that a laser is a solution looking for a problem. It was true that the uses of a laser were not immediately known to these young inventors. In the years to come, countless uses of lasers appeared. Here is a partial list of the way lasers are used today:

Laser surgery
Laser pointers
Laser aiming devices
Laser levels
Laser guidance systems
Laser CD and DVD players
Laser speed detection

This list is very finite and many more applications exist. Scientists and inventors alike looked at the properties of lasers and were inspired by their unique position in the world of science.

THE PROPERTIES OF LIGHT AND LASERS

Light Waves

Like radio and radar, light is comprised of a series of waves in a given second. These waves have the same relationship as radio waves in that the higher the frequency, the shorter the wavelengths. Light is much higher in frequency than radio waves and therefore has very short wavelengths. It is important to recognize that there are different kinds of light. When light is mentioned to most people, the first thought is of visible light such as the sun or a flashlight. But visible light is composed of all the colors of light and each color has its own wavelength and frequency. If you were to place a prism in the path of visible light, the waves passing through the prism glass will bend (refract) according to the wavelength of each color, which is why you can see each individual color of light. Also, there are other nonvisible wavelengths. On the lower end of the spectrum you will find infrared and on the higher end you will find ultraviolet.

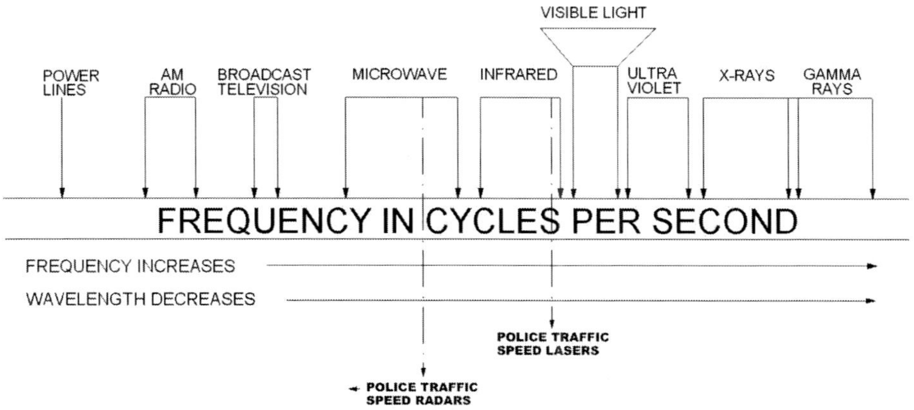

Figure 13. The electromagnetic spectrum.

In Figure 13, you can see the relationship of the different waves. Notice that visible light occupies a small portion of the spectrum when dealing with light. Because the frequency of a wave in light is very high, the proper way to describe light is by its wavelengths instead of its frequency. In K-Band radar, for example, the device is said to operate at 24.150 gigahertz; a measure of frequency, not wavelength. The description of light, however, is the reverse; red light starts at 750 THz (terahertz) and would be written as 750,000,000,000,000. The wavelength of red light is about 620 billionths (nano) of a meter. This may be a difficult concept because of the inverse relationship of wave frequency to wavelengths. Table 7 shows the individual frequency for light.

Table 7.
COLOR FREQUECY RANGES

Color	*Frequency Range*	*Wavelength Range*
VIOLET	668–789 THz	380–450 nm
BLUE	631–668 THz	450–475 nm
CYAN	606–630 THz	476–495 nm
GREEN	526–606 THz	495–570 nm
YELLOW	508–526 THz	470–490 nm
ORANGE	484–508 THz	590–620 nm
RED	400–484 THz	620–750 nm

The use of wavelengths as opposed to wave frequency to describe light is better for those who calculate equations using light. If there are those of you who enjoy such mathematical exercises, I'm sure you will agree.

Light Speed

We have already discussed this concept as it relates to radio waves, but when working with laser speed detection, you will find the knowledge of this property to be of great value. As previously stated, light travels at a constant speed of 186,282 miles per second (300,000 kilometers per second). Even though this is an incredible speed, it is a measurable speed. In astronomy, the knowledge of the speed of light allows us to determine great distances. The average time it takes light to travel from the sun to the earth is 8 and 1/3rd minutes. By multiplying this number by the speed of light, we can determine the average distance. The average distance from the earth to the sun,[1] for example, is measured to be about 150 million kilometers (one astronomical unit or 1 au). During the Apollo lunar missions, NASA determined the distance from the earth to the moon using a laser and the speed of light calculation. This was important to the determination of the amount of fuel and oxygen that would be needed for the trip.

Reflection, Refraction, and Absorption

Also like radio waves, light travels forever in a straight line unless reflected, refracted, or absorbed. With light, reflection is the way we are able to see things. The light waves bounce off almost everything, giving us a look at our world. In fact, the concept of invisibility would be a matter of simply bending light around an object. If the light did not reflect, then we could not see it.

Light also bends based on the density of the material it is passing through. Take a pencil and place it in a glass of water. The pencil seems to split in half. In reality, the light is bent by the water due to its density, and gives the impression that the pencil is in two places at once. With radio waves, refraction is a concept we must visualize in our minds. With light we can see what refraction looks like. In Figure

1. This number is an average since the orbit of the earth is somewhat elliptical.

Figure 14. An example of light refraction.

14, the young girl's arm appears to split as she reaches into the fish tank. As you look through the top of the tank, her arm appears to be suspended at a different angle than when you look at it from the glass side of the tank.

And for absorption, most light on the planet earth is absorbed by water in the air. This can readily be seen when clouds shade the sun. All of these properties are shared by light and radio waves.

Hazardous Waves

There is another property of light that we can observe. This property is the tendency of light to gain hazardous characteristics as the frequency increases. If we look back at Figure 13 and examine the light on either side of the electromagnetic spectrum, we can see the hazardous effects. Below red (the first color in the spectrum) we see *infrared.* The waves in infrared light have longer wavelengths, and a lower frequency. To humans, infrared light is generally harmless. But, on the other end of the visible light spectrum is *ultraviolet.* The waves in ultraviolet light are shorter and exist at a higher frequency. It is the ultraviolet rays of the sun that can burn human skin. A quick look up the spectrum and we can see other harmful forms of electromagnetic waves such as X-Rays and Gamma Rays. Of course, power and exposure play a role in how damaging these waves are, but there is no denying that there are hazardous effects from shorter wave lengthlight waves.

What Is Laser Light?

So what is a laser (or more specifically, what is laser light)? A television remote control is often an infrared (IR) devise, and so is a traffic laser; so is the television remote a laser? To best understand what a laser is, we need to look at the acronym LASER. The word stands for **L**ight **A**mplification by **S**timulated **E**missions of **R**adiation. It was coined by Gordon Gould as a description of the process by which laser light is created. An important note is that laser light does not occur naturally and must be the result of several components placed together. There are some additional properties that set laser light apart from ordinary light.

Coherent light are light waves that are "in phase" with one another. For example, two waves are coherent if the peaks of one wave are aligned with the peaks of the other and the valleys of one wave are aligned with the valleys of the other. Light produced by lasers are coherent light. Light from light bulbs or the sun would not be considered coherent. Laser light is often described as having waves that are collimated, which means the same thing. It is this property that helps laser light spread very slowly. Many think of a laser beam as a pencil thin beam of light and that it never spreads out. This is not true. All laser light will spread out wider; it is a natural tendency of light.

When laser light was first demonstrated, it was a sharp red color. This was due to the ruby rod that was used to create the light. Lasers do have the property of being monochromatic, of one color. But since the first laser, methods have been used to produce laser beams that encompass several colors which, with the help of a prism, can be separated into several beams of different colors. Some of these lasers,

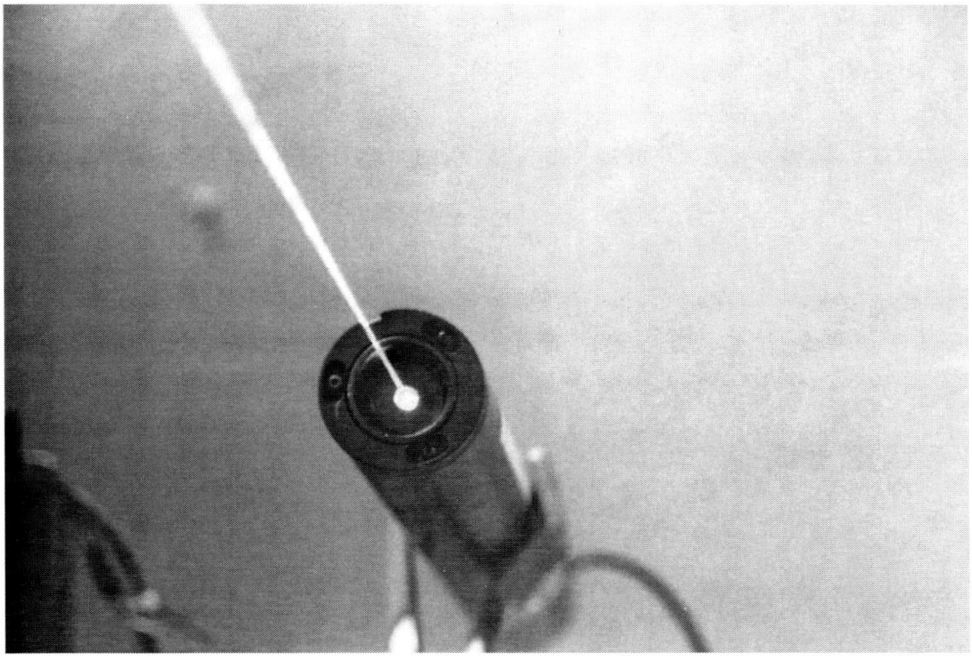

Figure 15. A helium-neon laser beam.

which are used for entertainment, produce a laser beam that has a white appearance to it. It can clearly be said that even these lasers occupy a narrow portion of the light spectrum. In Figure 15, we see a Helium-Neon laser beam. The gas is in a glass tube encased by the metal housing.

We have all seen the science fiction show or movie where the laser cuts through steel or simply blows up whatever it's pointed at. This property is, in part, true for a laser beam depending on some design considerations. What is common to all lasers is that they have high intensity. This is not to be confused with heat, although heat may be a part of this property in some cases.

Let's look at a good example of what we mean by intensity. If you were to get the opportunity to lay on the beach anywhere in the tropics, there would be a point when your skin would become irritated by the sun (sunburn).Without sun block, this may occur as soon as 20 or 30 minutes of constant sun exposure. Now, if you were to tape your eyelids open and stare at the sun for 20 to 30 minutes, well–this would be high intensity.

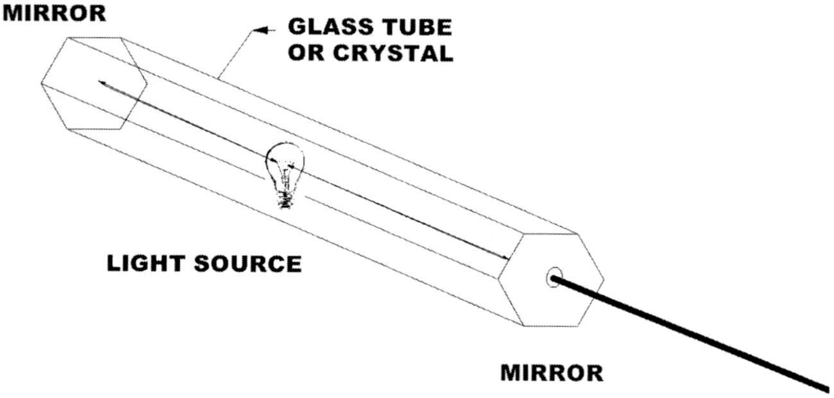

Figure 16. Line drawing of a traditional laser.

If you think about the acronym "LASER" as a sentence, you can see that it would read, "Using a LIGHT source and by AMPLIFYING this light by STIMULATING the EMISSION of its RADIATION (light) producing a coherent beam." Next we will break this sentence down into the method used to create a laser beam. This process is not very complicated, but there is a degree of precision required.

For our explanation, use the wire diagram provided in Figure 16 to help you envision the process. The first step is to obtain a source of light. For our discussion and a consistent understanding, we will use a light bulb to represent the source of light, although in most of the lasers manufactured today, a diode is very commonplace. By providing energy to the light bulb, it will shine (emit) light in all directions. The next step is to place two mirrors directly opposite each other and at a set distance. This will cause the light to reflect back to the filament. When the light waves reach the filament, they will "tickle" the atoms causing the emission of more light. Then the light travels on to the next mirror and this process is repeated. Each pass of light over the filament creates the emissions of light to intensify. The waves become coherent causing the light that is bouncing between the mirrors to become more concentrated. One of the mirrors is made thinner than the other and placed at the end where you will want the laser bean to pass through. Once the light becomes of sufficient intensity that it can pass through the thin mirror, the process of stimulated radiation ends and a laser beam is produced from this process.

Lasers require a medium or a substance that they pass through. The first laser was created by shining a high-power flash lamp on a ruby rod with silver-coated surfaces acting as the mirrors. Since then, there have been many substances that have replaced the ruby rod.

Gases have been very common. For example, helium and neon gasses will create red laser light while Argon gas lasers will produce a blue laser light. These lasers generally are a glass cylinder encased in a metal housing to protect them. As time and technology advanced, it was discovered that the use of diodes (silicon glass) was a very economical way to create low-power lasers. Applications such as laser pointers, and CD/DVD players capitalized on this method, and considering the other products now using the laser diode, manufacturing produces these diodes in the hundreds of thousands daily.

SPEED DETECTION LASERS

The common speed detection laser takes advantage of the gallium-arsenide diode which produces light in the infrared part of the spectrum commonly referred to as "near-red". It operates at approximately 904 nm wavelength. This operating wavelength is dependent on the specific diode used, but it will be close to 904 nm. Several diodes set into an array are used in these lasers.

Calculating Distance with Lasers

One of the early uses of a laser was the discovery that a laser beam can be used for very precise measurements. Today, laser survey equipment commonly called a theodolite is commonplace. The method used to take these measurements is a function of time and distance. In our earlier chapters, we discussed how we can determine speed using time and distance. In this case, the speed is our unknown and time and distance are known. This was done by using the distance over time equation:

$$\text{Speed} = \text{Distance} / \text{Time}$$

With a laser used to measure distance, the unknown and knowns are changed. In this case, we know the speed of laser light, and we can determine the time the light travels. What we are seeking is the distance it traveled. Using algebra, this changes our formula to:

$$\text{Distance} = \text{Speed} \times \text{Time}$$

In the case of a laser, the speed is 186,282 miles per second (300,000 kilometers per second).

To make this work, the laser is designed to pulse. To understand a light pulse, think of water coming from a garden hose. If you attach a nozzle to the hose to control the water flow, you can turn the water on for a brief moment and then off. This causes the water to have a definitive beginning and ending point or a period in which it exists. In the same way, light can be turned on and off very quickly creating a pulse of light that is akin to a pulse of water in appearance. The description of converting the speed of light into an expression of feet per second (English) or meters per second (Metric) can be found in Table 8.

Table 8.
HOW TO MEASURE WITH LIGHT (IN ENLGISH AND METRIC)

ENGLISH	METRIC
Each pulse of light from the laser will travel at 186,282 mps, so if we know how long the pulse travels, we can determine how far it has traveled as well. For example, if we emit a pulse of light for 1 second, it will travel 186,282 miles. If the pulse travels for one-half second, it will have traveled 93,141 miles. Of course, these are tremendous distances. Since light travels so fast, the time periods must be very short. If we were to convert the speed of light from miles per second to feet per second, then we would find that light travels 983,568,960 feet in one second. Now let's round this number to 1 billion feet per second. Of course, the laser does not round the number but rather works with the actual speed in feet per second. For this book, however, it is easier to understand these processes if we round this number to an even billion.	Sending a pulse of light from the laser, this pulse will travel at 300,000 km/s; so if we know how long the pulse travels, we can determine how far it has traveled as well. For example, if we emit a pulse of light for 1 second, it will travel 300,000 km. If the pulse travels for one-half second, it will have traveled 150,000 km. Of course, these are tremendous distances. Since light travels so fast, the time periods must be very short. If we were to convert the speed of light from kilometers per second to meters per second, then we would find that light travels 300,000,000 meters in one second. For consistency, we need to express this number as a portion of a billion, so 300 million meters is 3/10ths of a billion meters.

If we take the time of one second and divide it by 1 billion, we have this second broken down into billionths of a second or *nanoseconds.* This number is now small enough for us to take short distance measurements with a laser. For U.S. calculations, light travels 1 billion feet in one second or 1 foot every nanosecond. In metric, light travels 3 million meters in one second or .3 meters every nanosecond.

Let's look at a practical example. Pointing a laser transmitter at a wall, we emit a single pulse of laser light towards that wall. The laser pulse travels to the wall and is reflected back to the laser. A timing circuit determines the total travel time in nanoseconds. If that travel time was 151 nanoseconds round trip, then the pulse traveled 151 feet (46 meters). Of course, this pulse traveled round trip, so the laser must divide by 2 to get the actual distance, which in this case, would put the wall 75.5 feet (23 meters) from the laser. We are speaking in rounded numbers, but the laser is very precise in this determination. In laser surveying equipment, the distances can be to the fourth decimal place or greater. In speed laser, the precision will be, at best, tenths of a foot or 3 cm.

Calculating Speeds with Lasers

If we were to measure the distance to the wall as in our example above, wait a period of time, and then measure it again, we should get two identical readings for the distance. That is because the wall has not changed its position with relation to the laser. Of course, if we take a reading and then move the laser closer to the wall and take another reading, we can logically say that movement with relation to the laser was detected. Likewise, if we take two distance measurements in a period of time, and get two different distances, and the laser has not moved, then the object we are taking the distance readings from must be in motion with relation to the laser. The identification of relative motion in the last logical conclusion leads us to the fact that if we know the distance the object moves and the time it took for the object to move, we can determine the speed of the object.

Speed lasers utilize the distance divided by time calculation to determine the speed of a vehicle. By taking a distance reading for a vehicle, and then after a known period of time, taking a second reading for that same vehicle, the laser can calculate the amount of change in distance that the vehicle moved, giving a solution to the time-distance-speed calculation.

Example:

A vehicle is in motion with relation to the laser. The laser emits a pulse of laser light that travels round trip to the vehicle in 800 ns. The laser divides the time in half and places the vehicle at a distance of 400 feet (120.1 m). After a period of one-third of a second, a second pulse is emitted and travels round trip to the vehicle in 756 ns. Dividing by 2, the laser determines a distance of 378 feet (113.4 m). The laser then calculates a change in distance of 22 feet (6.7 m). With knowledge that the vehicle move this distance in one-third second, the laser can determine that the vehicle is moving 66 feet per second (20 meters per second). This information is then converted to mph (km/h) and the result is 45 mph (72 km/h). The math would look something like this:

Distance 1 (d1)

(U.S.)	*(Metric)*
Pulse travel time = 800 ns	Pulse travel time = 800 ns
Distance 1 (d1) = 800 ns / 2	Distance 1 (d1) = 800 ns / 2
Distance 1 (d1) = 400 ns	Distance 1 (d1) = 400 ns
Distance 1 (d1) = 400 x 1 foot	Distance 1 (d1) = 400 x .3 meters
Distance 1 (d1) = 400 feet	Distance 1 (d1) = 120 meters

Distance 2 (d2)

Pulse travel time = 756 ns	Pulse travel time = 756 ns
Distance 1 (d1) = 756 ns / 2	Distance 1 (d1) = 756 ns / 2
Distance 1 (d1) = 378 ns	Distance 1 (d1) = 378 ns
Distance 1 (d1) = 378 x 1 foot	Distance 1 (d1) = 378 x .3 meters
Distance 1 (d1) = 378 feet	Distance 1 (d1) = 113.4 meters

Change in Distance (Δd)

(U.S.)	*(Metric)*
Δd = d1-d2	**Δ**d = d1-d2
Δd = 400-378	**Δ**d = 120.1-113.4
Δd = 22	**Δ**d = 6.7

Determining Velocity (v)

Velocity (v) = Change in distance (**Δ**d) x 3

(U.S.)	*(Metric)*
v = 22 x 3	v = 6.7 x 3
v = 66 fps	v = 20.1 m/s

Determining Speed (s)

s = v / (5280/3600)	s = v x 3600
s = 66 / (5280/3600)	s = 20.1 x 3600
s = 66 / 1.4666	s = 20.1 x 3600
s = 45 mph	s = 72.3 km/h

This process may appear to be very involved, but it is actually very basic. But there is a little something extra the laser determines: speed lasers can tell direction of travel. Unlike radar, the laser in our example finds the second distance closer (smaller) than the first. This results in a determination of an approaching vehicle. If the change in distance had been greater, then the vehicle would have been moving away from the laser. Most lasers show direction of travel in their display. This information can be useful in verification of the identity of the targeted vehicle.

In radar, the determination of speed is the result of a change in radio frequencies. In laser speed detection, the process is very dependent on timing and distances. It could be said that lasers are a time and distance device, but, as you can see from the math we just covered, all distances are measured by time determinations also. Therefore, the description for what principle the laser is based in would be the *time of flight* principle.

We have given our example based on two pulses of laser light. In reality, the laser for speed enforcement uses over 100 pulses of laser light. These pulses are emitted automatically in series. In one second, a laser can send between 100 and 300 pulses depending on the model. As we stated, only two pulses can detect motion and could determine speed. But it is not possible to hold the laser steady enough to prevent the potential of errors in the calculation. As the operator holds the laser, there can be slight changes in the targeting point that can make a difference in the speed. For example, if the laser pulse was to reflect from the vehicle grill, and the second pulse reflects from the vehicle windshield, the change in linear distance would be viewed by the laser as additional motion. If the distance from one targeting point to another was as small as 3 feet (1 meter), the speed determination could be as much as 6 mph (10 km/h) higher or lower than the actual speed. That would be based on two laser pulses. However, speed lasers base the speed on many pulses compared to each other. For instance, if the laser emits 120 pulses in a second, then a minimum one-third of a second is required by the laser to determine the speed of a moving object. With 120 pulses, a total of 40 pulses would be used to make that determination. If in one-third of a second the laser was unable to get a consistent speed based on 40 pulses, it would keep trying until it was able to get at least 40 consistent returned pulses.

The mathematical process by which the laser determines that there is a valid speed reading is called the *average of least squares,* a mathematical process commonly used in regression calculations in statistics. The operation is quite involved and mathematically sound. To actually go into a further explanation of the least squares process would take a chapter all its own. But we can describe its function in somewhat lay terms.

Each time the laser determines a speed, it enters the speed in a register, a kind of place where a list of numbers can be held. The first speed is entered and no other action is taken. The next number will be compared to the last number entered to see if the numbers are consistent. The list may look like this:

45, 45, 46, 45, 44, 44, 47, 45, 45, and so on.

The laser will consider this information as good information as long as the speeds are a good fit. If the speed calculated is too different than the speeds collected, it will be ignored.

For our example, we used with the laser reflecting from two different targeting points, consider that the first 25 hits on the vehicle were taken from the windshield, and then there was a shift. The next several hits are on the grill. The laser has a limit to what it will accept as speed data, and the comparison of the 25 pulses returned from the windshield would be inconsistent with the new pulses from the grill. The laser would regard the data as a jump in speed too far away from the previous information. The laser would then seek to establish a new data list. This would continue until one-third of a second's worth of data was collected and a speed could be displayed. I realize the name of this process implies that the data is simply squared and averaged, but it's not as simple as taking these numbers and averaging them together. That process allows for too many potential variations in the final outcome. In this manner, the laser sees a list of speeds that looks like 45 will be the result. But the operator moves the laser to another targeting point. There is a jump in speed and the data is ignored. Finally the laser settles on this new targeting point. Since all parts of the vehicle are traveling the same speed, 45 again becomes the consistent reading. The data that was collected as the laser was changing targeting point has been eliminated from the outcome of the laser's determination of a speed.

I was told once that lasers do not make mistakes. Such a statement is a challenge to some. Since I felt challenged, I decided to explore the limitations of the laser. Since I was intricately involved in developing laser training, I was given a laser to use. One day I was using the laser on a major highway. I was monitoring a large semi when I observed a sports car behind the semi traveling faster. I quickly moved the laser from the front of the semi to the sports car. As I did, I noticed the speed that I had been monitoring the semi at increase by 12 mph (19 km/h). This not only got my attention but concerned me as well. I started to experiment with the laser and discovered that what had happened was that the laser added the movement of changing the targeting points to the speed of the semi. This also was supposed to be eliminated through, what the manufacturers called, error trappings or the least squares process. So, why did this happen?

Through study of the laser and experimentation, I determined that the laser is set by the manufacturers to eliminate jumps in speed that are outside a limit that is set at the factory. This is a function of the average of least squares process. I found that a gradual change, or a "sweep," can create an allowable jump in the speed reading as long as it stays inside these limits. This is called sweep errors. To understand this concept, we will take the speeds obtained by the laser pulses and isolate the distance portion. As a vehicle moves towards the laser, the distances are getting smaller (see Figure 17). Normally, if the targeting point is changed and a sudden change in the distance is experienced, the laser ignores the data until a new point is established. Once the new point is established, this new data will give a speed reading consistent with the true speed of the target. In Figure 17, the descending line is the established speed from hits on the windshield of the vehicle. The lower line represents new hits on the grill. The lack of data was the time that the laser was moving and information could not be validated. Note that both lines are the same angle, indicating the same speed.

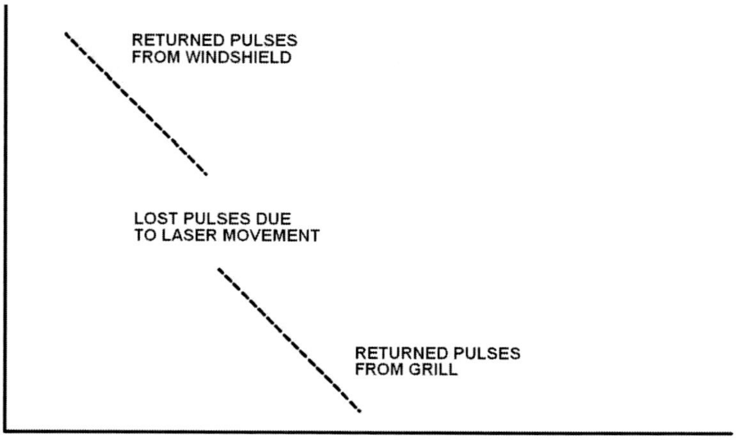

Figure 17. A graphic representation of the Average of Least Squares.

In the situation I created, the target point was changed gradually. As I moved the laser from the front of the semi to the sports car, I was sweeping the side of the truck. As I did so, the data never exceeded the limits set by the manufacturer. Remember that the laser is checking for data in one third of a second of time, so it can update itself rapidly. In each third of a second, the newly acquired speed would cause the limits to change with it. Consider that the movement of my laser was 12 mph (19 km/h) over about 2.6 seconds. This would mean that the speed was increased eight times. If the motion was uniform, the result would be a speed increase of 1.5 mph (2.4 km/h) each time the data was verified. If the manufacturer set the limits to only allow a change of 2 mph (3.2 km/h) each third of a second (hypothetically), each change would be allowed and the speed of 12 mph (19km/h) would be added to the speed of the semi at the end of 2.6 seconds of motion. Figure 18 shows the result of a gradual change in targeting points which allows the laser to read the motion of the laser into the motion of the target vehicle. This can result in either a higher or a lower speed reading.

When a target is moving towards the laser and the laser sweeps towards the target, a higher than actual speed can occur. The error trappings in the average of least squares are set by the manufacturer to allow some movement. If not, the user would find the speed laser most annoying because any small movement would result in no reading at

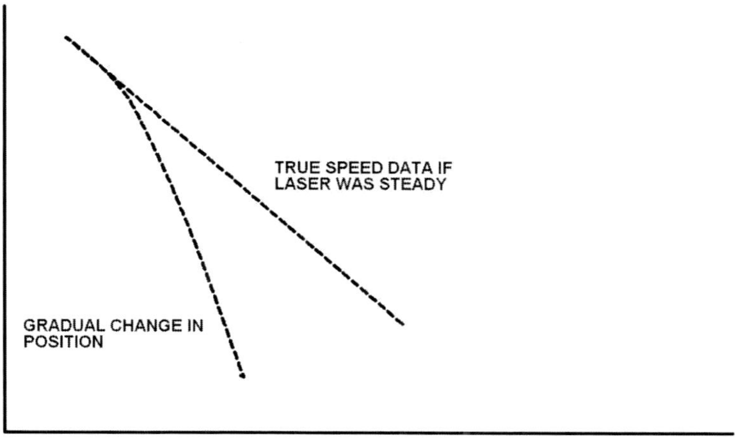

Figure 18. A graphic representation of a sweep error.

all. The operator must be continually aware of the potential for sweep errors and identify when they occur.

I was able to detect this error because I was using a laser that continuously updated the speed of the target vehicle. This is not to imply that there are lasers that would not sweep, but a laser designed to acquire a speed and display it without further updates can experience sweep errors and it may be more difficult to detect. The problem was that I was sending laser pulses while changing targeting points. Had I released the trigger to the laser while changing targeting points, then the sweep error I saw would never have occurred. Also, consider that the limits set for error trappings are different for each manufacturer depending on what they decide. After the sweep errors were discovered, these limits were made more restrictive to reduce the possibility of obtaining erroneous speed information.

Characteristics of Speed Lasers

So let's talk a bit about some of the characteristics that make up a speed laser. You must consider that the engineers saw a benefit to using lasers in speed enforcement. After they built their speed lasers, what characteristics does it have? Following are some of the characteristics we can identify.

Beam Character: The laser beam is created by several diodes. These diodes are passed through a lens system which creates the final laser beam. This beam does diverge over a distance. Most speed lasers increase in size at the rate of 2.8 feet (.85 m) for every 1,000 feet (305 m) of linear distance. The common operational area of a laser beam (the area most operators will target a vehicle) on open highways is a distance of 800 to 1,200 feet (244 to 365 m). A great deal of hype is associated with the maximum distance a laser can determine a moving object. I have observed a laser indicating motion with an associated speed at a distance of 5,500 feet (1,676 m). I don't know what the target was, but it did give the reading. The operator must be able to identify the target when obtaining a speed reading, so the limitation of the speed laser is the operator's ability to see the target and not the range of the laser. The beam is not exactly cone shaped but more oval or even square. At 5,500 feet (1,676 m), the laser beam is 15.4 feet (4.7 m) wide. If the targeting point is the center of a lane, this will allow the laser to spill over into another lane.

Reflective Character: Lasers detect returning pulses from the most reflective target point. If an operator targets an approaching vehicle headlight on the driver's side of the vehicle, depending on the distance, the laser may elect a vehicle moving away from the laser by picking up a taillight. Most vehicle taillights are designed as a retroreflective lens, allowing for nearly 100 percent of the light hitting it to be returned. In this case, angle plays a very minor role in the determinattion of which targeting point to select. The intensity of the returned light is the largest factor. Color and paint condition (shiny or dull) also have a role in the laser's ability to see the returning light. Dark colors absorb light, while lighter colors reflect. Shiny paint also returns more light than dull paint. This is not to say that the laser will not see a dull black car, but the laser will take a little longer to see a dull black car, than a shiny white car.

Pulse Identification: A laser can become confused as to how long a pulse has been gone, but such a condition happens almost exclusively with very reflective objects. Remember that the laser is counting the number of pulses, looking for the equivalent of one-third of a second's worth before displaying a speed. If something interrupts the pulse return, the laser can make a brief mistake on the distance the pulse determined. While this situation may occur when the laser is measuring a distance, the error is always trapped by the average of least squared process. Each pulse travels from the laser to the target and back before the next pulse is sent. A laser with a pulse repetition rate of 300 will emit a pulse 300 times a second or once every 1/300th of a second (.03 seconds). If the pulse is to travel to a target 6,000 feet (1,829 m) away, the pulse will be gone 12,000 ns (.000012 seconds) round trip. This is approximately 2500 times faster than the interval of time between two pulses. So the laser will notice if a pulse does not return, and it cannot mistake one pulse from another. What can happen is, when there are mirror-like objects, the pulse can be bounced from the mirror surface to another surface and then returned. This would delay the return and give an improper distance reading.

Overexposure: The laser can see the returning light pulse even over most other light sources. Headlights, for example, cannot eliminate the laser pulse. Although headlights, to us, seem very bright, they consist of white light, which is the entire spectrum. Lasers are looking for their particular wavelength, and theirs is very bright. Pointing the laser at extremely bright objects like the sun, however, may cause the laser to lose some of the return pulses. If the laser cannot locate a pulse, then no reading is produced. Operation of the laser in a westerly direction at sunset can cause difficulty in obtaining speed reading due to the laser losing some of the emitted light pulses.

When the laser itself was invented, scientists marveled at what they had created. Laser light does not occur naturally anywhere on earth. (I know I am deflating some of you who are Indiana Jones fans, but, no, the staff of Ra does not shoot a laser beam at the final resting place of the Ark.) But when it was invented, the scientists at some point must have asked, "so, what do we do with it?"

Chapter 8

SPEED LASER OPERATION AND TESTING

When the laser determines a target speed, it will give the operator three pieces of information. First, it will give a speed reading. If the laser is designed to give a single readout and not update, the speed will be at the final point of calculation. If the laser is designed to constantly update, then the reading will change in concert with any changes in speed the target vehicle makes. The direction of travel of the vehicle is also provided. If the target vehicle is moving away from the laser, a minus sign will appear in the display. If the target vehicle is moving towards the laser, there will either be no minus sign or a plus sign will appear. Finally the laser will report the distance the vehicle was at when the speed reading was determined.

OPERATOR CONFIDENCE

You may have noticed that we did not cover tracking history yet. Actually, we will not be identifying a "tracking history" for lasers. When the first speed lasers were sold, the marketing approach was that tracking history was not needed. It was the intent of the manufacturer, Laser Technologies, Inc, to overcome the burden of target identification that existed with radar. This would eliminate the need for a tracking history, but there is still a burden of proof requirement on the part of an accuser. Since we are not speaking of automated speed enforcement (that's yet to come), the accuser is the operator of the laser speed detection device. With that in mind, the following is provided to establish this burden of proof. It is justifiable to differentiate tracking history from laser as the technology is more target specific than radar and as such eliminates most of those burdens.

The speed laser operator needs to take certain steps to have a high confidence that the accused has violated the speed laws. In some cases, these steps have semblance to tracking history and in some cases they are completely new.

Visual Determination of a Speed: it is imperative that the operator expresses an allegation that a law was violated. When that allegation is that a vehicle was being operated in excess of a legal speed limit, the operator will need to testify to the speed that it was believed the vehicle was traveling. This precludes that there will be an estimation of the speed of the vehicle. In speed laser operation, this is more natural since the operator will select which vehicle to target. Generally, that selection will be based on an estimate.

Acquisition Tone: a sound that the laser produces which lets the operator know that the target has been acquired. This is NOT a Doppler tone in most lasers. This tone identifies good reception of returning laser light. The clearer the tone, the more light is being received. The exception to this explanation is found in the Stalker lasers. The Applied Concepts, Inc. manufacturer allows their lasers to give a pitch which would correlate to a speed. The laser is not making a Doppler shift determination, however. Instead, the Stalker laser associates the pitch with the speed found in the display. The stalker laser still gives the operator information that the laser is receiving a strong return of laser light, or lack of acquisition. The pitch is an added feature to help the operator use both eyes and ears to determine target identification.

Correlated Speed Readout: so that the operator sees that the speed estimate made agrees with the speed displayed by the laser.

Direction of Travel: this information will solidify that the operator did not mistakenly get a vehicle traveling in the opposite direction.

This process will establish the violation alleged. Notice that we did not include the distance information. It is very difficult to estimate distances, even for law enforcement officers. To say "yep, that looks like 341 feet (or 100 meters) to me!" is unreasonable. The distance information is provided to locate where on the road the vehicle was when the speed reading was taken and this information can be useful. For example, you may wish to use the laser to measure the distance from your position to the start of a school zone. Say that distance is 600 feet (or 200 meters) from you and the laser then gives a distance of 570 feet (190 meters) on a target vehicle, you have the confidence that the violation occurred inside the school zone.

LASER ANOMALIES AND JAMMING

Lasers do not have as many anomalies as radar. Even in radar, most anomalies are in moving radar. Still, there is one worth mentioning. Like stationary radar, the laser can be subject to cosine. Exact distance measurements are based on straight line measurements. Movement of a vehicle from point "A" to point "B" on a road is accurate if the measurement was taken along the path of the vehicle. But the laser measures at angles to that path, since the operator is situated off the roadway (hopefully). That makes these measurements subject to the same cosine that radar experiences. Also, like stationary radar, this cosine is in favor of the moving vehicle in that the speed will always be lowered.

For the most part, there are not many opportunities for radio signals to interfere with laser light. Although I have not seen any source of interference for lasers, all speed lasers sold in the United States are required to have a detection device that will alert the operator that there is interference in the area. Nonetheless, people have attempted to create laser interference devices in order to prevent the laser from acquiring a target vehicle or otherwise to jam a laser. One of the more popular attempts is to flash an infrared light at the laser or to return laser pulses to confuse the laser. The main problem with this method is that the device designed to create the interference must operate at the same pulse repetition rate as the speed laser. Since every laser manufacturer sets this rate, and the rate sometimes changes for different lasers made by the same manufacturer, the potential for successful interference by this method is the same as winning the lottery.

OPERATIONAL CONSIDERATIONS

Using the laser is pretty straightforward. Nonetheless, there are some things to be considered when operating your laser. Foremost is to hold the laser steady. Radar is like a shotgun. You can shoot from the hip and you will still probably hit something. Laser is like shooting a .22 rifle. Steady sighting is required to obtain a speed reading. I watched potential customers at a traffic safety products trade show pick up a laser and try pointing it. They were not familiar with aiming the laser and target acquisitions took eight to 10 seconds. One of the problems is that they did not use proper stance.

Steadiness

Part of the consideration for how you hold the laser will be the targeting elements, or sights. Some lasers have actual scopes to look through. Others have a heads up display, or HUD. And still others have a camera view finder-type window. The transition through these came from modifications in design and from user input. Each targeting element uses a slightly different design. Most of them are a dot or circle to pinpoint the location you wish to place the center of the laser beam. Early lasers had a shoulder stock which aided in holding the laser steady. Tucking the stock into the shoulder like a rifle helps greatly in keeping the laser steady. However, if you couple the shoulder stock with a scope, it appears that you're shooting at the motorists. I recall a woman who drove past a speed laser and then called the supervisor to complain that the police were shooting at her. This design moved to a laser that could be held without a stock. Today, the majority of lasers are designed to be handheld. There are several models that are more like a camera in design as well. Lasers with a Heads Up Display (HUD) can be sighted using either one hand (Figure 19) or two hands (Figure 20). Also, in examining these figures you will note that the officer in Figure 19 is using both eyes open, while the officer in Figure 20 only uses one eye to sight with. The heads up display affords the operator the ability to do either.

One of the more important considerations with the sighting ability of the laser is that the laser targeting elements are mounted on the roof of the device. This means that the laser sights and the laser beam do not begin to coincide for a particular distance. If the laser sights are two inches higher that the transmitting lens of the laser, then at a distance of 1 foot (one-third meter) from the end of the sight, the laser beam will still be lower that the sighting element. At the factory, the

Figure 19. Single hand method.

Figure 20. Two hand method.

lasers are bore sighted. Since the laser beam is diverging slowly, the manufacturer sets the sighting element in line with the center of the beam at a particular distance based on the expected range of the laser. The manufacturer will determine what distance the laser sighting element (dot or circle) is at the laser beam center. Generally, because of that setting, lasers will not detect a speed at any distance less than a certain distance; speed readings will not display fewer than 25 or 50 feet for example. On the far end of the beam, the sighting element will remain inside the laser beam longer because the laser beam is spreading. In the operational range of speed lasers, the targeting element remains inside the laser beam.[1]

In laser operation you will gain a better ability to hold the laser steady over time and use, but you also will be moving the laser from target vehicle to target vehicle. This runs the risk of getting sweep errors. Through experience, you will gain a better ability to hold the laser steady. Utilizing proper technique, you can determine if the laser result is correct. However, it is best to avoid sweep errors altogether if the reading of the laser is correct, but it is even better to avoid experiencing sweep errors. The best way to do this is to release the trigger and allow the laser to clear before engaging the next target vehicle. In the situation I gave with the movement from the semi to the car, it would have been a simple step to release the trigger of the laser when I

1. Some states set the maximum operational range of the speed laser so there is no question as to proper sight alignment. Florida, for example, sets the maximum enforcement range of lasers to 2,000 feet or less (Florida Administrative Code 15B, 2011).

moved from one vehicle to another. Develop the practice of releasing the trigger each time you move the laser from one vehicle to another.

Also, be sure your laser has a clear field to the target. With radar, objects in the foreground may be of little importance, but with lasers, they can make the difference in obtaining a speed reading. I was operating a laser in a construction zone and after only a few minutes, I found I could not get a speed reading on any target. It was obvious that the laser was trying to acquire a target based on the tone, but it never could finish. Looking at the area between myself and the laser I noticed evenly spaced barricades. I realized that the laser could not complete the speed data, because I would pass the laser over each barricade, causing the laser's error trapping software to reject the data. Also, in night operation, objects that you do not see can be on the way. Telephone poles and signs may not be illuminated and will interfere with the laser's ability to acquire a target vehicle.

Safe Handling

Lasers are not toys, they are tools. Like any tool, improper use can get someone hurt. It is very important not to point the laser into the eyes of a person. I know the logical question; "how can I point it at a vehicle and not at the driver?" To answer that question we will need to discuss the eye safety of lasers.

Lasers are set into one of several categories of eye safety. They are: Class I, Class II, Class IIIa, Class IIIb, and Class IV. Of these classifications, only Class I is considered eye safe. These classifications are based on several factors such as power output and laser wavelength. Lasers do not have to be big or expensive to be dangerous. Most common laser pointers are Class IIIa, meaning they can harm the human eye. This part of the human body is used for this standard due to the fact that the human eye is very light sensitive. The optic nerve is also nonregenerating and damage to this nerve is permanent. The agency in the United States for designating the laser classification is the Center for Devices and Radiological Health of the United States Food and Drug Administration. In Europe, the lasers meet similar standards as regulated by the International Electrotechnical Commission.

Speed lasers are classified as Class I based on their output beam, the resulting light emitted after the lens system. If the lens is cracked or removed, then the laser classification can be different. Looking directly into the lens of a speed laser runs the potential of causing injury to the human eye. As you move away from the beam source, the intensity diminishes, and safety increases. In any case, care should be used not to point the laser directly at someone's eyes. The highest concentration of beam energy is at the center of the beam, so when targeting a vehicle, it is always best to avoid the windshield. Consider that at most distances there will be a degree of spillover to the driver's area during targeting, but targeting the headlight/grill areas of the vehicle diminishes the potential exposure. If you inadvertently target the driver's area, it's a good chance there will be no injury, but these cautions will help insure the driver's safety.

Keep in mind also that these devices are designed to be used on ground-operated vehicles only. NEVER POINT A SPEED LASER AT AN AIRCRAFT! In 2005, a man pointed a visible light laser at a commercial airliner on final approach to Teterboro Airport in New Jersey, causing the pilot to become distracted. Such a situation could end in an aircraft crashing. The man was arrested for his actions. I realize that some may believe this would not apply to speed lasers as they are infrared and cannot be readily seen, but some pilots (especially military pilots) use night vision equipment when operating aircraft. Most night vision equipment is designed to enhance what the user sees by using and intensifying natural infrared light. An infrared laser is incredibly bright to night vision equipment.

As you can see, the lens is very important to the safe operation of the laser. The operator should take care to see that the lens is not damaged. Some laser lenses are glass, while others may be made of acrylic. Most are treated for infrared operation. If the laser is dropped or struck, the lens should be inspected and if there is any visible abnormality, the entire unit should be sent for servicing. Cleaning the lens is actually very easy. If needed, wipe the lens with a moist cloth of soap and water, and then use a dry cloth to remove any excess moisture. Never use chemicals, lens cleaners, or chemically treated cloths on the lens. Also be aware that the lens will react to temperature and humidity like any other piece of glass. Moving the laser from an air conditioned vehicle to a warm and humid climate will cause the lens to fog. Also, if the laser lens has moisture on it and is placed in a sub-freezing climate, it may result in ice crystals. Always remember to change the environmental conditions gradually. Sudden or extreme temperature changes can crack the lens.

SPEED LASER TESTING

National and International Testing

Like radar, lasers undergo a series of tests for determining that they perform as specified. Germany is very active in the testing of lasers for European application. In the United States this testing is done by the International Association of Chiefs of Police (IACP). Testing is voluntary in the sense that if the manufacturer would like the laser available for purchase with federal tax dollars, then, that laser must be included on the IACP Conforming Products List (CPL). In general, the testing is as rigorous as the tests for radar, but there is an interesting difference. In radar, the technicians can send a radio signal to the radar to simulate a speed reading. For lasers, this could not be done. To provide a possible simulator for lasers, a device had to be built that could do the job. This laser simulator, built by Laser Technology, Inc., interfaces directly with the laser. Using this interface and a computer, the laser can be fooled into thinking it is detecting movement. Here's how it works:

Lasers have two lenses: a transmission lens and a receive lens. The transmission lens emits the laser pulses, and the receive lens collects them on return. The interface fits directly over the laser's lenses. A laser simulator is shown in Figure 21. Once connected, the attached computer is used to tell the laser something is moving. The computer is set to the tested laser's pulse repetition rate, and then a speed and distance is entered. The laser is then triggered. The emitted laser pulses travel into the simulator and disappear. The simulator then sends back its own laser pulses, causing delays to simulate movement. The result is that the laser will display speed, direction, and distance which should accurately duplicate what the technician programmed into the computer.

There was some controversy when this simulator was first used as some felt that Laser Technology, Inc. should not be involved in the construction of the simulator since the IACP would be using it to test Laser Technology's own speed lasers. This concern was quickly put to

Figure 21. A laser test simulator coupled to an LTI 20-20 laser.

rest when it was pointed out that someone had to build the simulator who knew the technology. Also, the electronics industry practices this method. When a piece of test equipment is calibrated (which is essential to proper testing), it is often calibrated using equipment made by the same manufacturer that built the test equipment being calibrated.

Daily Speed Laser Testing

Like radar, lasers need to be tested for proper operation both before any enforcement is taken, and after such enforcement occurs. These lasers need to be subjected to the following tests and pass each test according to the manufacturer's specification. Like radar, it is very important for the operator performing these tests to know each proper test requirement and the proper test result.

Automatic Power-Up Tests: The laser has a built-in test sequence for power-up. The exact information depends on the model, but at minimum, the laser will generally do a full light test and check its internal circuitry. The power-up tests should not be considered complete tests but regarded as abbreviated tests. The following series of tests should be conducted by the operator daily.

Light Test: This test is designed to assure that all the lights on the laser are working properly. This includes the speed display(s) and all warning lights. This test MUST be conducted first for a failure in any of the illuminated portions of the radar display could affect the results of other tests. Most lasers use liquid crystal display (LCD) in the display. Because of the high reliability of the LCD displays, failures are rare, but if there is a failure, the radar needs to be sent to an electronics repair facility. In any case, a failure of the light test would require the radar be removed from service until the problem is corrected.

Internal Test: There are a variety of internal tests depending of the manufacturer's design. Early lasers had an internal beep which was an audible method to alert the operator that all internal components were functioning properly. Since speed lasers do not function on the Doppler principle, there is no need for the laser to verify any Doppler shifts. Speed lasers use timing as a key component to determine both distance and speed so the internal tests generally check the timing circuit and the logic ability of the laser processor. At times, a simple "Pass" is displayed or for some lasers, the absence of an error indication is all that will be needed to pass. For the internal test, it is imperative that the operator is aware of which indications (or lack of indications) would be required. The laser MUST display the exact information required for this test. Failure of the internal test would require the laser be removed from service until the problem is corrected.

Sight Alignment Test: This test is a method to make sure the laser beam is located where the targeting element is placed. The manufacturer recommends the distance that the laser should be checked. If the laser has an internal sight alignment check function, then the procedure described by the manufacturer should be followed. If there is no function in the laser for sight alignment, then the operator can target a small object such as a reflector post that is about three to six inches wide (7.5 to 15 cm) at a distance of 250 feet (76 m) or greater and take distance readings both on and off the post. If, while the sighting element is on the post, the readings are the correct distance, and while off the post, the distance readings are not equivalent to the post distance, the test will have passed.

This test needs to be conducted on both the "x" and "y" axis (horizontally and vertically). To accomplish this, the laser would be checked in the normal operational position, and then turned on one side and checked again. Also, consideration must be given to the fact that most speed laser beams are not round. If examined through an infrared camera, the laser beam would appear oval or somewhat rectangular in shape at close range. To compensate for this, you may wish to set a second target at a different distance and offset by one-half the laser beam width. In this manner, the laser could be pointed between the two targets and see neither target.

Distance Test: Again, some laser manufacturers have procedures for this test. However if they do not, then the operator can simply target two objects that are at a known distance and verified by the operator (the operator has personal knowledge of the distances because the operator measured the distance with a tape measure), and take measurements with the laser. The resulting distance readings should be within 1 unit of measure (1 foot or 1 meter) accuracy. Since the calculations of a laser are based on the time of flight of the laser pulses, the measurement determines that the laser's internal clock is correct.

Delta Distance Test: Some manufacturers require a Delta Distance Test. This particular test is similar to the Distance Test, but tests the ability of the laser to make a change in distance determination. It is important to understand that the Distance Test checks the laser's ability to determine distances. In some lasers, the distance determinations are accomplished with a different average of least squares process. When a laser measures a distance, it is expected to be static or nonmoving. So why employ the average of least squares process since there was no change? The Delta Distance Test allows the laser to determine a change in distance of a target while in simulated speed mode. This test follows these steps:

Reading 1: Take a distance reading of a known distance. This distance must be over 50 feet (15.25 meters) away from the laser.

Reading 2: Take a distance reading of a second target at a known distance. This distance, too, must be at 50 feet (15.25 meters) or more and must be a different distance that the first. The laser will display both distances as they are taken, and then will calculate the difference between the two. If the two distances are separated by 30 feet or 10 meters, the laser will show that distance.

Calculation: The next step will be for the laser to calculate a speed based on a constant of 1 mile per hour (or 1 km per each 1/2 meter), so the laser will display 60 mph (based on the 30 feet above) or 20 km/h (based on the 20 meters above). Also, if the farther distance was determined by the laser before the shorter, the laser will indicate a target moving towards the laser. Conversely, if the closer target was determined before the farther one, then the laser will indicate a negative number as a target moving away from the laser.

These tests need to be completed by each operator twice daily. The legal reasons will be discussed later in the book, but the desire for a determination that the equipment is operating correctly is allowed testimony as it goes to the desire that there is no error when determining the speed of a target. Safe and proper laser operation will maintain both highway safety and a trust by the public in those efforts by law enforcement.

Chapter 9

LEGAL CHALLENGES TO SPEED DETECTION

THREE KEYS TO THE LEGAL USE OF SPEED DETECTION DEVICES

Over the years, people have approached me and asked how to beat the speed detection device. My guess is that they believe that there is some trick or loophole that I would know and they want me to share it. Legal council has tried to hire me as an expert witness in order to give them weight in a case. My rejection of doing so is based in my belief that speed detection is a sound and reliable device, and a suspicion that my involvement is designed to "muddy the water." In one case from a neighboring agency, I had a defendant contact me and ask how he could get a copy of the training manuals for radar. In my discussion with him, I discovered that this defendant was not interested in justice, but rather in finding a loophole to use to get off. I told him how to buy the books, and then I called the officer in the case and warned him of what the defendant was attempting. I knew the officer and had trained him, and I was confident in his ability. On the day of trial, I was in the courtroom on a case of my own and I was able to see how this case was resolved. The defendant walked to the stand with the training manual under his arm, but by the conclusion of the case it was apparent he never read it. The verdict: guilty!

From the inception of speed detection, there have been challenges. I have learned that some do not like being told they are doing something wrong when driving. There is an emotional tie to the driver, and to be accused is damning to him/her. Once, I had a speedometer clock on a vehicle at 120 mph (192 km/h). When I told the driver how fast he was going, he replied with a determined voice, "I wasn't going over 110."

With the advent of radar, and now laser, people have brought to the appellate level of courts claims that the radar or laser speed information was not valid. We will look at the three issues, or "keys," that will lock down these devices as valid. Understand that these keys must be used. In other words, these devices can be misread by the operator. The challenges we are going to speak of set a foundation to how speed measurement devices must be used.

Key One–Scientific Principles

The first question to be asked by the legal system is whether or not the device operates under sound scientific principles. I was in another state watching the news when a story came on about a guy who had invented a crime finder device. This device had an antenna and would search like a divining rod. At some point, it would stop and point at someone. The inventor stated that when that happened, the person it was pointing at had committed a crime. In the news story, he tried to explain that it picks up human reactions of guilt without making physical contact. When the reporter tried to pin him down on how it operated, he stated, "well, it works under a physics principle that only I understand." You can see the folly in this, but the real scary part was that the news media found a police department that had one.

State of New Jersey, v. Dominick Dantonio ((June 20, 1955) 18 NJ 570, 115 A3d 35, ALR2d 460.)

On February 2, 1954, Dominick Dantonio was operating a bus on the New Jersey Turnpike when he was stopped by state police for speeding. He was cited for speeding 66 mph (105 km/h) in a posted 60 mph (96km/h) speed limit. The device used to determine his speed was a radar unit. Dantonio's bus had a tachograph device mounted in it, a device that prints out a graph of the speeds of the bus at various times. This became a battle of technologies for the court. In this case, the court ruled that the principles of radar, specifically the Doppler Principle, were widely accepted by the scientific community when used to determine speeds through frequency shifts. This case established judicial notice of the Doppler Principle as the scientific basis for radar operation. The outcome of this case was that the conviction was upheld by the court.

Admissibility of Motor Vehicle Speed Readings with Lasers, Morris County Superior Court of New Jersey, Law Division ((June 13, 1996)314 N.J.Super 211 [714 A.2d 371])

Forty-six years later we find ourselves back in New Jersey on a decision affecting laser speed detection. In this case, several defendants were charged with speeding and were heard in court. In all cases, the speeds were determined using a Marksman 20-20 laser by Laser Technology, Inc. This was a global hearing into all aspects of laser speed enforcement. In this case, the court stated, "After considering all of the proofs and the analysis and the argument presented, I am satisfied that the general concepts of using lasers to measure is widely accepted in the relevant scientific communities and is valid." This established that the determination that laser speed enforcement was a valid method of speed detection and that judicial notice could be taken of this method.

With these two cases as examples, we have established the methods used to determine speeds are valid scientific concepts. Here we are speaking to the methods of radar in the use of Doppler Principle, and in lasers the use of laser beams in a time-distance calculation of speed utilizing Time-of-Flight methods.

Key Two–Testing of the Device

One thing commonly confused is a statement that the court can take judicial notice of the device used. This is somewhat untrue. These are instances where this statement can be used as it pertains to a specific radar or laser for a specific period of time. For example, if an officer testifies that at 9:00 a.m. he tested his radar, and at 3:00 p.m., he tested it again, and through that testimony he described those tests, then the court can take judicial notice if the testing of that radar for all cases heard between those times. But the court cannot say that all Decatur Genesis II, Select radars are valid at all times. Testing is required on those devices because they are an instrument, and such instruments can have part failures due to wear and tear.

State of Connecticut v. Tomanelli (1966) 152 Conn 365, 216 A2d 625.

This case is most notable to mention for radar testing and probably one of the most misrepresented cases. In this case, Michael Tomanelli was stopped for traveling 64 mph (102 km/h) in a posted 45 mph (72 km/h) speed limit. At trial, Mr. Tomanelli admitted to speeding, but claimed his speed was reasonable and prudent. In his appeal, the defendant contended that he had raised the issue of the accuracy of the radar device and indicated that the judge had taken judicial notice of the radar. The Connecticut Supreme Court established a much stronger position on the judicial notice of the scientific principles of radar than in Dantonio. In its ruling, the Connecticut Supreme Court stated, "There can be no doubt, at this late date, of the general scientific acceptance of the Doppler-shift principle upon which police radar operates. As explained by the Austrian scientist Christian Johann Doppler over a century ago and as utilized by instruments such as the one used in the present case, the difference in frequency between the wave emitted by the radar set and the echo wave can, if the instrument itself is accurate and is properly operated, be translated into a velocity reading of extreme accuracy."

The court went on to identify that judicial notice cannot attach to specific devices and testing is required by stating, "Judicial notice can extend, however, only to the scientific accuracy of the Doppler-shift principle as a means of measuring speed if the principle is correctly applied. Judicial notice does not extend to the accuracy or efficiency of any given instrument designed to employ the principle." In short, testing would be needed to prove that the device is working properly.

In this case, the radar used was an analog-designed unit that showed the speed of a target vehicle by a speedometer-type needle. Also, graph paper could be added to the device to allow for a hard copy record of speeds. The device was designed by the manufacturer to be tested regularly utilizing a series of tuning forks. By striking these tuning forks and holding them in front of the antenna, the radar's needle would jump to the indicated speed. By using these tuning forks to test the accuracy of the radar, the court stated, "Under these circumstances the accuracy of the radar unit was unimpeached."

At times when I have heard this case discussed, the word "calibrate" would come up. With the radar used in this case, that word was valid, as well as for several others that followed. Calibration implies to correct something, or to make one device match another. With the radar used in this case, the needle could lose some accuracy and require adjustments. A friction shoe behind the panel can get worn and the operator could get an incorrect reading. To overcome this, the radar had a dial marked "calibrate" and the operator could turn this dial while holding the tuning fork, moving the needle to match. For example, if the radar showed 57 when the tuning fork was designed for 60, a simple turn of the dial and the needle could be moved to 60. It is not mentioned in the case brief, but I believe that the calibration feature must have been discussed in the court and it may be why the court ruled so strongly concerning tuning forks. In fact, by the late 1970s, the word "calibration" was commonly used to describe radar testing.

With the advent of digital technology, tuning forks hold a very small place in the testing processes. Canada, for example, has no precedent case laws concerning tuning forks. They are not held to United States court decisions, and as such have no laws requiring the use of tuning forks. However, if the manufacturer indicates they must be used, then they have to follow the manufacturer's guidelines.

State v. Reck (Dec. 21, 1994), 2d Dist. No. 1352 CA, 1994 Ohio App

In this 1994 case, two prevailing decisions were established:

1. The device used must have been tested for proper operation before enforcement activity, and
2. The operator must have received sufficient training and experience in laser speed detection.

The first decision shows that, like radar, lasers must be tested in accordance with the methods prescribed by the manufacturer. Notice that the courts do not attempt to specify the proper testing in lasers as was the case in Tomanelli. It is left to the manufacturers to specify what tests must be conducted.

Key Three–Operator Training and Experience

How much training is needed to run radar or laser? When they were first introduced, the manufacturers would send their sales personnel to train the operators. Of course, while training, they are not selling. So are they giving complete training? A deputy sheriff in Georgia told a reporter that he had received a 45 minute lecture from the company, and he was now getting on-the-job training by using the device. The State of Florida still requires 40 hours of radar operator training, which is made up of 24 hours of classroom and 16 hours of field work. Let's take a look at the cases that indicated such training is needed.

Honeycutt v. Commonwealth of Kentucky ((November 4, 1966) [408 SW2d 421])

In this case, Erna Honeycutt was stopped for traveling 50 mph (80 km/h) in a posted 35 mph (56 km/h) speed limit and was cited. The speed was determined by similar radar to the one in the previous Tomanelli case. In the request for appeal, the defendant claimed that the officer did not have sufficient training and experience to interpret the speed information given by the radar. The court advised that ". . . it is sufficient to qualify the operator that he have such knowledge and training as enables him to properly set up, test, and read the instrument; it is not required that he understand the scientific principles of radar or be able to explain its internal workings; a few hours' instruction normally should be enough to qualify an operator."

This concept of setup, test, and read has prevailed over the years, but the training was expended a bit with the advent of moving radar. With that, a second case is important to address.

State of Wisconsin v. Hanson ((1978) 85 Wis.2d 233, 270 N.W.2d 212)

Lawrence Hanson was stopped for traveling 68 mph (108 km/h) in a posted 55 mph (88 km/h) speed limit. In this case, the officer testified that he was using a Kustom MR-7 Moving Radar in the moving mode of operation. There were a multitude of questions in this case, but the findings were strongly held that moving radar operated utilizing the same (Doppler) principles as stationary radar so that the judicial notice of stationary radar applies to moving radar. Furthermore, the court held that one hour of training in moving radar and three months of experience is sufficient under the law.

It is interesting to note that the defendant had succeeded in having the charge dismissed by the local court. It was the state that appealed this decision and obtained a reversal. The second trial ended in a conviction.

State of Florida v. Aquilera (1979) (no citation)–Case Docket # 711-0105 Dade County Florida

So if an hour or less of training is sufficient, where did the idea of up to 40 hours of training come from? In 1979, a Dade County (Miami) court took 43 pending speeding cases and asked for evidence that the devices used were valid for speed enforcement. After a lengthy and nationally reported trial, all 43 cases were dismissed. From this court case, both the National Highway Traffic Safety Administration (NHTSA) and the State of Florida examined law enforcement training in speed enforcement. The outcome was training recommendations that ranged from one to two weeks. However, these were just recommendations.

In 1994, with the case of *Reck v. Ohio,* and in 1996 with the Morris County, New Jersey case, the courts remain with a concept of "sufficient" training and experience for laser speed enforcement. Exactly what is meant by sufficient is not clear, and it is felt that the courts wish to leave the definition to the local courts. In some cases, states like Florida and North Carolina have mandated training. Others leave the hours to what the local courts will accept. The prescribed training program by NHTSA today offers a combined curriculum for Radar and Laser Operation with a recommendation of 24 hours classroom and a suggested 16 hours in the field for a total of 40 hours. The Northwestern Center for Public Safety offers 40 hours of Radar and Laser Instructor training. Clearly the hours of training are defined by case law to mean whatever it takes to bring an operator to a level of proficiency in order to be accepted by the court of jurisdiction.

I am sure that early laws that regulated speeds were disliked. There is always two sides to any question, and both sides seem passionate about this one. Our goal may be to get public acceptance and compliance; however, there will always be someone unhappy when there is an issue with traffic laws.

Chapter 10

SPEED ENFORCEMENT METHODS

Everybody has their idea of how speed enforcement is to be conducted. From completely passive to very aggressive, there is a way to enforce speed laws. In this chapter, we are going to examine the more popular methods and concepts used throughout the world. The intent is to get you thinking about both how and why we use a particular method of speed enforcement.

You will find throughout this chapter that I believe drivers needs to be given fair warning. As humans, we are creatures of habit. Some habits are good and some are bad. But we tend to rely on our habits to get us through the day. Traffic violations are often a result of a habit and if there was just a little bit of reminder available, then the behavior may change without causing a negative impact financially or physically. Of course, there are those who have a problem with getting the warning. Such people tend to lose my sympathy sometimes. So let's move through this chapter from passive enforcement techniques to the best active enforcement methods.

PASSIVE SPEED WARNING SYSTEMS

Passive speed warning systems can be broken into two groups: traffic calming and traffic warning. The word passive is used here to indicate the lack of hard enforcement by a summons or other charging document. The systems are intended to call the drivers' attention to how fast they are traveling or to make them somewhat uncomfortable with their speed.

Traffic Calming Techniques

If you have been involved in law enforcement for any length of time, you have or will have someone complain of the speeds of vehicles on their street. I know that I had to, on occasion, send my officers to some lonely street where there were probably 10 cars a day driving by, in an attempt to try to catch a speeder that some citizen believed cursed their neighborhood. I don't mean to make light of the complaint, but I feel for the officer who sits all day waiting for this demon to drive by. I am also aware of the opposite situation, where the violations are so numerous that the neighborhood needs its own officer. Placing an officer for long term in either of these situations is not a good use of staffing.

One remedy is to design the road so that speeding is not desired. The application of a redesigned roadway can afford relief to the problem. The citizen commonly wants the jurisdiction to place a stop sign on the road to curb the speed. This design is not the best resolution to the problem. A stop sign sets up an unexpected annoyance to the driver. It tends to switch the problem from speeding to stop sign violations. In fact, in some cases, the stop sign can create a traffic crash problem. Consider the fact the motorist wants to drive fast and then has to stop. This often results in a desire to make up time on the other side of the stop sign. Additionally, as the stop sign becomes routine, then the motorists start to ignore the sign completely. A stop sign is designed for two things: to allow cross traffic an opportunity to enter the roadway and to give breaks in the traffic flow. Often the placement of a stop sign to reduce speed has neither of these. The result is a trust in the sign by those respecting it in contrast to physical interaction (a crash) with those who ignore the sign.

Traffic calming is an adaptive way to make the motorist feel like exceeding the speed limit is unreasonable. Methods include;

Speed Humps: a speed hump is a rise in the road for a period of time, generally about a car length. This is not a speed bump, which is a short rise in the road. The disadvantages to speed bumps are that they place a strain on the vehicle which can cause damage, and that they tend to cause loss of control at higher speeds, placing the agency liable for them. Speed humps on the other hand have a longer duration so they cause less damage and it is less likely the driver will lose control.

Mini-Roundabouts: you have seen these used often in Europe, at times in larger forms. The mini-roundabout creates a traffic control without the need for a stop, yet causes the driver to change direction and therefore automatically slows the driver in order to negotiate the roundabout.

Offsets: if there is sufficient land, then the road can be given a series of offsets or jogs, the best is to have two in a row. These offsets need to be sharp, which is allowed in low speed areas, in order to make it difficult for the driver to try and negotiate through the turn.

There are a variety of other techniques that can be used. These solutions require an effort in concert with the law enforcement agency and the traffic engineer to produce. Obviously they are best used in the initial development of a neighborhood, or in the revitalization of a part of the community. Their application is costly and time consuming. With the time and money necessary, this is not a quick fix to the problem.

Traffic Speed Warning Systems

If time and money are a concern, then the application of a traffic speed warning system can be used. Traffic speed warning systems are displays that alert drivers to their current speed, and give warning when they are going over the posted limit. The system may be simplistic in design such as a pair of yellow lights that flash when an approaching vehicle is traveling too fast. I have seen these warning systems throughout the United States and in some other countries. Generally, they are set in an area where speeding or speed-related crashes are prevalent. As a vehicle approaches this warning, a detection system (either radar or pneumatic hoses) identifies that the vehicle is exceeding a preset safe speed and the lights are activated for a short period of time. Generally, this system is mounted permanently at a location.

The more contemporary approach is speed displays or messaging signs. Some of these are like the flashing yellow lights in that they are permanently mounted at a particular location. Most others are mounted on a trailer that can be conveniently towed to any needed location (see Figure 22). These speed displays operate by using directional sensing radar so that only the approaching traffic speed is displayed. These speed displays can cost between $2,000 and $20,000 (U.S.) depending

Figure 22. Speed warning sign in rural location.

on the options they contain. Some are designed to show just the approaching traffic speed, while others show both the approaching traffic speed and the posted speed limit.

The more advanced type of speed warning trailers (and the more expensive) are the variable messaging trailers. These systems display from one to four lines of information. They can be used to remind motorists to drive safely. They can also be used to advise motorists of unusual road conditions ahead. Such a system is shown in Figure 23. Generally, these systems are mixed with a radar that will intermittently (or continuously if needed) display approaching speeds. Depending on the sophistication, these systems can be costly. Also these system messages are regulated by the Department of Transportation when used on state and federal highways. Your DOT has a list of approved message ("John just turned 55" will not be one of them).

With a portable system such as this, care must be taken for proper placement. A speed display can become a crash cause instead of a

Figure 23. A variable messaging sign in use.

crash remedy. The display needs to be located so that the motorist can see the numbers at a sufficient distance to allow for safe deceleration. Never place a display in a sharp curve or at an intersection. This may cause the motorist to slow too quickly or otherwise distract the motorist from other vehicles. Also, many of these systems have an over limit warning such as red flashing numbers, or flashing lights. The point of the sign is to get the motorist to slow down to the limit, so utilizing these warnings can help in this regard. Also, most of these signs are manufactured with good "see through" visibility in mind. Nonetheless, you need to be sure the sign is not blocking a crosswalk or driveway.

One of the major advantages to the system is the fact that it can be placed in one location and moved to another. In this manner, it's there one day, and gone the next. Motorists become complacent with seeing the same thing every day, so mobility of the system keeps them watchful. The chosen location should be an area of needed enforcement. Many of these systems have an option to attach a data collector to them. A good technique is to place the system for a week or two at a suspected location in need of speed reduction, but either leave the speed display off or the sign down. While in this state, have the speeds of traffic passing collected by the data collector and analyzed. After this, turn the sign to its operational state, and conduct a second study to see the effectiveness of the sign. You could also conduct a third study by collecting data with the sign off, after you have used it for a period, to see how long the driver behavior remained in compliance.

A major consideration to any speed display is how much it will be used. Some of that usage is dependent on the power supply. The system may run on batteries that need periodic recharging, or they may run on solar power. But once charged, having the system sit at the back lot of the station until needed is not a good use of the dollars spent. There are locations that the system can always be used at, such as school zones, in front of schools and parks, and highway construction sites. The city I worked for had 24 school crossings. Placing the system in each crossing for a week would take the entire school year to give total exposure.

AGGRESSIVE ENFORCEMENT

The point is to get people to want to follow the speed limits, and the preceding techniques can be very effective. But, at times, people need motivation. The United States Transportation Research Board issued results of several studies that were conducted after the national 55 mph speed limit was eliminated. They reported a 5 percent increase in the 85th percentile speeds where the limit changed from 55 to 65, and a 25 percent increase in fatal traffic crashes in urban interstates, and 18 percent increase in fatal crashes in rural interstates.[1] There are a great many considerations to these statistics, but one that would be clear is the decrease in enforcement that would have occurred. Since the results indicate that the 85th percentile was skewed down 5 percent, then enforcement efforts would have been greater to

1. "Managing Speed" (Special Report 254) 1998, Transportation Research Board.

keep compliance. Once the limit moved towards the speeds people wanted to drive, it would be logical to believe that less enforcement would be needed. However the de-emphasis would cause law enforcement to shift their presence.

Enforcement Methods

As I have alluded to previously in this book, enforcement is a state of mind. A motorist sees a police officer on the side of the road with a motorist. The reality could be one of many things: a lost motorist seeking directions, a broken-down vehicle, or a warning being issued. But to the passing motorist, it is a citation being issued. Presence is 90 percent of the mission to enforcement of traffic laws. If the public thinks law enforcement is in the area, their behavior changes. Ominous presence has a great impact, but what people think about the law enforcement agency's mission to traffic enforcement is even greater. The important part of what people think is to get them to buy into the program. As a teenager, there was a community near my hometown that had a reputation for strict enforcement. My friends and I would avoid that community for fear of being stopped for just about anything. This is NOT the desired approach to speed enforcement.

People must know that you are serious about safe driving, but they should not be afraid. People are pretty smart. They know what radar is, and they know why you may be driving a slick top patrol car. Seeing them is an advisement not to speed. As a police sergeant, I drove an unmarked patrol car and I used both radar and laser from time to time. I do not believe that I caught more people with an unmarked car than I did driving a marked one. I do believe I had more interesting violations I would not have gotten otherwise. The concept is not to count the violations. Numbers does not impact the problem. Judgment and discretion on the part of the officer should never be subject to micro-management. Neither should unbridled speed enforcement be a part of the routine. It is very easy to fall into the pit of seeking opportunities to simply "write tickets." To grab a radar or laser, and go to the favorite spot and just stop some speeders is the fastest way to public ill will. Effective speed enforcement is a blend of the officer's daily contacts with the public. Every warning has weight as well as every citation issued. If the approach to speed enforcement by the agency is that "speed enforcement is everybody's job," then the tools needed must allow for the officer to remain mobile. Exposure throughout the patrol zone will do more good that sitting in one spot all the time. Don't get me wrong; sitting in that spot has its purpose and use, but it should not dominate the patrol activities. You will go farther with your efforts by making your day consist of contacts with people and talking with them, applying the appropriate enforcement to the situation, than spending the day issuing 25 or 50 citations. Patrol with a purpose, not a ticket count.

Concentrated Speed Details

There are times when certain speed details are needed. Warnings hold no weight when there are no teeth behind them. This approach should not be arbitrary. A speed or crash study should support any concentrated effort. Let's walk through the uniform process of setting up such a detail. Remember, the goal is public voluntary compliance because they believe in the mission, and not because you have placed the fear of the citation in them.

First, determine a need. This can be done from three main sources:

1. From an evident or perceived crash situation: I had a lieutenant once, whose philosophy was that two incidents reveal a pattern. I agree with this in part. Two incidents can alert you that there is a possible problem. Even one incident, if severe enough, can create this suspicion. This will point you into researching the potential to see if a response is needed. Also consider the severity of the crashes. Sometimes the crashes can't be eliminated, but injuries can be reduced.

2. From citizen complaints: The residents may be bringing you the problem. However, most will not bring the complaint to you; you will need to seek it from them. Go to community meetings, and public events. Also, most people are reluctant to make the statement formally. If the event allows you to mingle with the people, you will gain more information than you could at any formal meeting. One of my favorite ways was to go to the local mom and pop restaurants to eat. People always enjoyed passing along their concerns there. A good friend of mine, Sgt. Joe Spaulding of the Clay County Sheriff's Office in northeast Florida, would eat lunch with me, and there was never a time he did not have to stop and speak with someone at any restaurant we ate at. Clay County has a population over 190,000 people, and Joe has his hand on the pulse of that community.

3. From observed violations: Be very careful in your study based on this situation. It is very easy for a law enforcement perception to be misunderstood. A clear study of the 85th percentile and correlated information to the potential of future crashes is vital to this method.

Next, visit the local traffic engineer. This person is a valuable asset to your efforts. Review your hypothesis and study information that supports it. See if that person agrees with your finding. They may have some insights to the cause. It is entirely possible that the engineer may be working on the same problems you are. Both of you working on the same problem will build a strong program.

Then, involve the media or otherwise advertise. You are selling a product, safety, and you want your customers to buy it. Tell the media what you are doing and why; especially why! The media loves to do stories on what government is doing, because the public wants to know. Also, consider having the media present for the detail. Your objective is to get the longest effect from the action you take, and the media will reach many more people than those who just drive by. Outside the media you can use ad campaigns, and even pass out informational flyers concerning where you are going to conduct the speed enforcement. At the beginning of each school year, I directed my motorcycle officers to go to the local elementary schools. The first week of school, the moms and dads were lined up at the pick-up lane waiting for their children. My motorcycle officers would ride up to each vehicle and give the driver a brochure on driving safety focused on school zones and child restraints. Now let's select the approach we will use to execute this concentrated effort.

The Special Detail

I basically do not like the term "Taxi Stand Operation," but it does describe the method. Personally I liked "Signal 15"; our code for such an operation. In this approach, an officer operating the device to calculate speed will be the base unit. This officer will monitor traffic and when a violation is detected, a chase vehicle will pursue[2] the violator. There are some rules to this operation:

---

2. Be sure to review your agency's policy on chases and pursuits. The words "chase" and "pursue" are used here to describe a typical attempt to stop a traffic violator. Be prepared for your responsibilities to a situation where a vehicle attempts to flee from police.

First is safety. The base unit generally will not be pulling into traffic, but the chase units will. Planning the entrance to traffic is essential to keeping the situation safe.

Communicate who the violator is and get confirmation from the chase unit. Possibly you are sitting next to each other, but maybe you are sending this information to the chase unit by radio. The best way to make sure the target is clearly identified is to have eye contact between the base unit and the chase units. In this manner, the ability to see the target aids in clear identification.

Make sure the chase unit stops the correct vehicle. Listen to the stop (if the stop is beyond your ability to see what is happening) and make sure the officer calls in a description that matches the vehicle you monitored.

Keep good records of each target that a chase car was sent to stop, and each stop that was made. Also remember to have the citing officer list the base officer on the witness list. All of this information will be important at any court hearing later.

The best location for the taxi stand operation is an interstate or rural highway. Taxi stand operations can be done with as many officers as deemed needed. If you consider the average time it takes for the base unit to clock a violator, send the chase unit out, the time to conduct the stop and return, and then divide that time by the number of officers on the detail, you can predict how many officers you will need. Keep in mind that other situations may remove that officer from the detail (call for service and arrests, for example). It is recommended to have a supervisor on site in this and all operations to deal with citizen questions and concerns and to liaison for the media that may be present.

The checkpoint operation is a reverse of the taxi stand. In the checkpoint operation, the target vehicles are driving to you. The real disadvantage to a checkpoint is found in the safety aspect. The target vehicle is driving to you, and a determination of speed is made. The next step is to attempt to get the vehicle to stop and pull off the road. Generally, the location for the stop should have an associated parking lot to pull the vehicle into. There are different ways to get the vehicle to slow and stop. The best method would be to use traffic cones on either side of the lane before the stop location. The driver will slow on the approach and then can be signaled off the road. Also a law enforcement vehicle on the side of the road or in the median can activate the overhead lights to warn the driver to slow and stop. In addition, an electronic messaging sign can be used to alert the driver to stop. In all cases, NEVER walk into the lane in which the target vehicle is traveling in order to make the stop.

Traffic control personnel (officers, reserve officers, police explorers, or police service aids) can be utilized to help control the situation. If it is obvious that the target vehicle is not going to stop, then a chase vehicle may be used to pursue and stop the vehicle after it passes through the checkpoint. The staffing requirements for this operation vary with the number of lanes involved and the volume of potential violators. At minimum, you will need a base unit to monitor and identify the violating vehicle. Next, someone would be needed as the communication/control unit to communicate directly with the base unit (when the number of enforcement officers gets larger, it is advantageous to have a single communicative point of contact to identify the violator and control the distribution of stops and information). You will also need units to control the traffic at the stop location, and units to take enforcement action. Finally, you will need a detail supervisor; someone who can talk with citizens or the media about what you're doing and to take any complaints that me be brought to your attention. Supervision can be cursory in most cases, but the detail supervisor needs to be available. This method of operation works well in business and some urban locations. It is not recommended to attempt this method on higher speed roads or interstates.

Airborne Speed Enforcement

Monitoring speed from the air is not new. "Airborne Speed Operation" is common with Highway Patrols and State Police organizations, which have used this method for years. We will look at this method as it has traditionally been conducted on open highways and interstates, and we will also examine how to employ this method in a city environment.

First of all, the speed detection method will be Time-Distance since it is not possible (yet, I'm sure) to run radar or laser speed enforcement from an aircraft. It is precluded due to the angles that result from the position of the aircraft to the ground. With this in mind, there needs to be a clear marking of the road for the distance portion of the process.

In interstate applications, this marking is a white line placed across the road. White is the only valid color to use since highways worldwide use colors to identify traffic control. Generally, there are two lines established at a measured distance. On highways where the limit is 60 mph (96 km/h) or higher, the lines are placed one quarter mile (1,320 feet) or one quarter kilometer (250 meters) apart. In an urban setting, the speed limits are lower, so the distance may also be reduced. This is especially important since there are very few stretches of road in a city with one quarter mile (or one quarter kilometer) distance without traffic lights. The point is to catch a speeding vehicle and it is not efficient to set your limits such that you cannot catch them. City applications may have the lines set at a distance of one city block or about 600 feet (182 meters).

Many places use the highway department to paint the lines on the roadway. The 3M corporation has a vinyl tape that is used for highway markings as well. This tape can be simply applied to the road surface and will stick to the asphalt. It is designed to be attached through a heat sink process, but this is optional, making the attachment very quick and easy. It is important to know where these stripes have been placed, so the first step before placing them is to measure and mark the distance. The location of the stripes can be marked using surveying tags and nails. Remember not to use a curve since curves are different lengths depending on your distance from the inside to the outside.

Once the lines have been established, the person in the aircraft will simply time vehicles as they pass over these lines. Once a violator is determined, the information is radioed to the ground.

The ground operation may be either in the form of the taxi stand or checkpoint. All that has changed is the location of the base unit which is now the aircraft. Some question as to the aircraft movement with respect to the movement of vehicles on the ground is always raised (generally in court). The most common aircraft for this method of speed monitoring is the Cessna 152 or Cessna 172. This is due to the high wing configuration of the plane. Figure 24 shows a Cessna 172 ready for preflight. Helicopters may be used and have been in the past. Generally this is due to the fact that the agency often already owns them. But this is not always the most economical approach. A Bell 47 (the M.A.S.H. type of helicopter) burns 17 gallons of fuel an

Figure 24. A Cessna 172 "Skyhawk" aircraft waiting on a speed enforcement mission.

hour. Even the Robinson R44 four-seat helicopter burns 15 gallons an hour. The Cessna 172 (this aircraft is better designed to carry two adults for two hours with total fuel) burns between 7.5 and 9.5 gallons an hour. At an altitude of 1,000 feet (304 meters) and a ground speed of 60 knots (69 mph or 110 km/h) the targets are very easy to monitor and make a speed estimate. Remember that on the ground, the targets are generally first observed at this distance, and in moving mode of radar at this speed.

In the interstate or open highway application, the aircraft determines a speed of a target vehicle and alerts the unit on the ground that begins to catch up to the vehicle. In this case, the aircraft stays with the vehicle until the ground chase vehicle is directly behind the target vehicle. The correct vehicle is confirmed, and the stop is made. The operator of the time distance calculation radios the time and speed (the distance is already known), and the aircraft returns for the next clock. In the city application, the process is similar. The aircraft determines the speed of the target and in this case notifies the communications officer who instructs the traffic control units to slow and stop the traffic. The target vehicle is caught up in the traffic situation, and the aircraft identifies which lane the target vehicle is in, traffic in that lane is moved forward until the target vehicle reaches the front of the line, where the aircraft identifies that vehicle and it is diverted into an enforcement location. If, in the city application, a vehicle fails to stop, a chase car is sent after the vehicle. Since the aircraft has constant sight of the vehicle, high speed is not required and the chase unit can simply be directed to the vehicle.

When I was at the Largo Police Department, I would decide on the date for our airborne speed enforcement detail (I dubbed "Operation Skyhawk") and then about four days before the detail, I would place signs on the road involved announcing "Speed Enforcement Ahead." The signs would then stay up two days after the detail. By placing the signs a period of time before the detail and then taking them down afterwards, I felt it would clue drivers in that they were going to be monitored. Since they didn't know when, they would have to start slowing down for the whole week. Hopefully, it would become a habit.

Automated Speed Enforcement

When I hear the term "Automated Speed Enforcement," I cannot help thinking of the scene in the motion picture "Demolition Man," where every time Sylvester Stallone uttered profanity, a nearby device issues him a ticket (the movie is recommended). I was of the full belief that it would not be successful since the public had such a strong opposition to it. But, once I learned how it worked and I saw it in action, I was pretty well sold on the idea.

What is it? How does it work?

There are basically four methods used to execute photo speed enforcement. The first one to emerge is the radar-based system. In simple terms, the radar-based system sends the radar beam *across the road,* instead of down the road. If you recall, this will create a cosine effect on the speed reading. In this case, the angle is known and the speed calculation can be corrected. By sending the radar beam across the road, the ability is established to capture only one target vehicle instead of a mass of vehicles. This allows the system to gather speed on only one vehicle, and by setting the timing for the photo, give the system pictures of both the front and the back of the target. I watched the system by Applied Computer Systems, Inc. (ACS) operate in Maryland (U.S.) and was very impressed with the technology.

Another method uses laser to determine the speed of target vehicles. In this system, the laser is scanning as opposed to a single beam. In this manner, it is possible to see multiple targets on the highway. Because the beam is still narrow, the system can calculate correct speed even in different lanes. It further has the ability to classify the vehicle tracked as to size and distance from the device. Like the across the road radar approach, the system knows the angle of the beam, and by proper placement, knows distances. The computer calculates the true target speed from the information it gathers. In Florida, I was able to use the system by Nestor Traffic Systems, Inc. and was convinced of its validity in determining speeds even in a multi-lane situation.

A company in Oregon, Hardin Optical, used a video camera and a form of time and distance to determine speeds. In this system, a computer determines the number of pixels used to create an image and as that image approaches or recedes, the changes in the number of pixels over time determines the speed of the vehicle. Because it is an image, by adjusting what you can see will single out a lane or vehicle.

Also using a camera, B & W Sensors of Missouri determines speed of the target through time and distance. Using the concept of perspective grid photography (actually a form of photogrametery), the system establishes geographic locations for everything in the camera view. In other words, it knows the distance from one point to another. By timing a vehicle between two known points, the system can assign a speed. Also, since the system knows the location of a broad range of points, it can assign speeds to multiple vehicles in multiple lanes simultaneously.

Many people are concerned about automated speed enforcement. They feel it's a money-gathering tool and has no highway safety value. After watching the system in Maryland, I am convinced it has a highway safety application. With the ACS system, we selected a residential road near the training classroom, and set up the radar. A sign was on the approach to the vehicle that the radar and cameras were mounted in. This sign stated "Photo Radar Enforcement." On the vehicle was a yellow flashing light and a radar display sign indicating "Your Speed." I was very skeptical concerning success. The system was set to capture speed at 15 mph (24 km/h) over the limit. In one hour, we caught 20 vehicles. The road was a low volume location for I am sure we would have caught more had the traffic been heavier. I was sold. The systems met my qualification that the action taken was valid as long as the motorist was effectively warned.

Nothing about automated speed enforcement can circumvent the principles in this book. The application must meet a need. And if the need is eliminated, then the system should not be used. The system should be a speed enforcement tool and the efficiency of eliminating law enforcement personnel should have no impact on the traffic safety goals of the jurisdiction. In this chapter, we placed a high importance on public acceptance of the way enforcement takes place. Automated speed enforcement, used properly, can fit that requirement.

Most companies do not charge for the systems. The system is given to the agency to use. They place it where it is needed and the system goes about its business photographing violations. These violations are then processed by the company which keeps a portion of the fine for operational costs and overhead, and sends a portion to the jurisdiction using the system. Generally, the fines levied are lower than actual speeding fines, and the violation is of a local ordinance instead of a state traffic law, so insurance companies do not raise the driver's rate. The perception can be that the system is a money-making venture. A common question of the company is whether or not they want to see the violations decrease, since the violations generate income for them. Hopefully, these companies do not view the process of their system in this fashion.

If the jurisdiction is looking at automated enforcement for speed laws, there are some considerations:

Do you have a need? Determine that there is a need for automated enforcement. An examination of your crash situation is first. What is the injury and fatality rate like; how many disabling damage crashes do you have and are these crashes located on roads that will benefit from automated enforcement? Can you use the systems to curb speeding in school and construction zones? Identify as many as possible to get a good picture and involve the city traffic engineer in these studies. And prioritize the problem areas.

What are all the tools at your disposal? Speed warning systems, staffing levels, and the probable commitment time to the tasks should be considered. Have these tools been effective in the past?

What is an acceptable resolution time frame? Of course, you want the problems solved as quickly as possible, but that may not be practical. Set goals for the resolution and then see if the system better fits the goals.

Once you have decided you are going to use automated enforcement, let the public know what you are doing and why. A clear program picture will assist you in your goals. If the system is portable, let them know where you plan to use it. Don't commit to when, so you can keep the motorist alert to changing their driving behavior. Most companies have a contract for the period of use so talk with neighboring agencies to see if they want to partner with you on the program. The big advantage to a portable system is that it can be shared. Use practical limits. What must be compensated for is that there is no visible officer on the side of the road; therefore, the other motorists are not aware they need to slow down. This can be compensated for through the advertising campaign, but you don't want overkill on the number of violations. Set the system for a speed enforcement level that matches the speeding problem in an area. You may want to start with citing for 20 mph (32 km/h) over the limit, and gradually lower the enforcement speed over time. Many of the companies offer the ability to issue warnings. Incorporating a warning letter into your program can be of great benefit. You may get a better response from drivers who get a warning letter than ones who get a citation.

SUMMARY

The point of this book is to give a law enforcement officer the most complete and up-to-date information concerning speed enforcement. It was not intended to be unabridged, but rather to be focused. I held a class in Canada once in which the students wanted to take the radars apart. I really did not have a problem with that; certainly to know your equipment that well speaks highly of what you can do with it. I know that once you start coming up with new ideas and creative ways to do things, you then have reached the level of expert. Don't get me wrong;

it's alright to reach a level of proficiency and be content with that, too. But I believe that something in your life should be a spark to being all you can be.

A good cop once told me that before you're done with your career, you will be crazy. Not crazy in the deficient sense; crazy in being different, being a person to solve other people's problems or prevent them from harm. So this book was written to balance the desire to excel and the desire to be content. This book was written on how to do speed enforcement correctly. Not with 200 tickets a month, and not with 200 warnings either.

The most important part of any speed method is to limit time and frequency. Certainly a ten-hour, five-day operation could virtually eliminate unlawful speed, but the outcome of distain for law enforcement will be far more detrimental than the success is worth. The overall goal is to get the public to buy in and voluntarily slow down. Combining the three "E's" of traffic safety will give you the best results. All the enforcement methods and tools addressed here are intended to be used in combination, with each other. Depending on what you're financially and logistically capable of, your program to curb speeding should be a global one.

BIBLIOGRAPHY

American heritage dictionary (2nd college ed.). (1991). Boston: Houghton-Mifflin.

Black's law dictionary (9th ed.). (2009). St. Paul, MN: Thomas West

Encyclopedia of science and technology (7th ed.). (1992). New York: McGraw-Hill. Vol 4. pp 362–363; Vol. 9, pp. 606–611; Vol. 10, pp. 39–41.

Fisher, P. David et al. (2002). *Defense of speeding, reckless driving, and vehicular homicide.* Newark, NJ: LexisNexis, Matthew Bender.

Giancoli, Douglas C. (2004). *Physics, principles with applications* (6th ed.). New York: Prentice-Hall.

Managing speed. Special Report 254. National Academy Press, Washington, DC, 1998.

Managing speed. (1998). Washington, DC: National Academy Press.

Olsen, Paul L. (1996). *Forensic aspects of driver perception and response.* Tucson, AZ: Lawyers and Judges Publishing.

Triola, Mario F. (2009) *Elementary statistics* (11th ed.). London: Addison-Wesley.

Vascar Plus Enforcement Program. (1976). Traffic Safety Systems, Richmond, Virginia.

INDEX